Salami

Salami
Practical Science and Processing Technology

Gerhard Feiner

AMSTERDAM • BOSTON • HEIDELBERG • LONDON
NEW YORK • OXFORD • PARIS • SAN DIEGO
SAN FRANCISCO • SINGAPORE • SYDNEY • TOKYO

Academic Press is an imprint of Elsevier

Academic Press is an imprint of Elsevier
125 London Wall, London EC2Y 5AS, United Kingdom
525 B Street, Suite 1800, San Diego, CA 92101-4495, United States
50 Hampshire Street, 5th Floor, Cambridge, MA 02139, United States
The Boulevard, Langford Lane, Kidlington, Oxford OX5 1GB, United Kingdom

Copyright © 2016 Elsevier Inc. All rights reserved.

No part of this publication may be reproduced or transmitted in any form or by any means, electronic or mechanical, including photocopying, recording, or any information storage and retrieval system, without permission in writing from the publisher. Details on how to seek permission, further information about the Publisher's permissions policies and our arrangements with organizations such as the Copyright Clearance Center and the Copyright Licensing Agency, can be found at our website: www.elsevier.com/permissions.

This book and the individual contributions contained in it are protected under copyright by the Publisher (other than as may be noted herein).

Notices
Knowledge and best practice in this field are constantly changing. As new research and experience broaden our understanding, changes in research methods, professional practices, or medical treatment may become necessary.

Practitioners and researchers may always rely on their own experience and knowledge in evaluating and using any information, methods, compounds, or experiments described herein. In using such information or methods they should be mindful of their own safety and the safety of others, including parties for whom they have a professional responsibility.

To the fullest extent of the law, neither the Publisher nor the authors, contributors, or editors, assume any liability for any injury and/or damage to persons or property as a matter of products liability, negligence or otherwise, or from any use or operation of any methods, products, instructions, or ideas contained in the material herein.

Library of Congress Cataloging-in-Publication Data
A catalog record for this book is available from the Library of Congress

British Library Cataloguing-in-Publication Data
A catalogue record for this book is available from the British Library

ISBN: 978-0-12-809598-0

For information on all Academic Press publications
visit our website at https://www.elsevier.com/

Publisher: Nikki Levy
Acquisition Editor: Patricia Osborn
Editorial Project Manager: Jaclyn Truesdell
Production Project Manager: Caroline Johnson
Designer: Matthew Limbert

Typeset by TNQ Books and Journals

Contents

Preface ix
Acknowledgments xi
Introduction xiii
Disclaimer xv

Part I
Meat and Fat

1. **Meat and Fat**
    - 1.1 Introduction — 3
    - 1.2 Amino Acids — 4
    - 1.3 Proteins — 6
    - 1.4 Collagen — 9
    - 1.5 Muscle Physiology — 10
    - 1.6 Flavor of Meat — 15
    - 1.7 Principles of Muscle Contraction and Relaxation — 17
    - 1.8 Enzymes in Meat — 19
    - 1.9 Fat — 20
    - 1.10 Rancidity of Fat — 27

2. **Biochemistry of Meat**
    - 2.1 Biochemical Processes in Meat Preslaughter — 31
        - 2.1.1 Glycolysis — 32
        - 2.1.2 Krebs Cycle — 33
        - 2.1.3 Oxidative Phosphorylation — 35
    - 2.2 Biochemical Processes in Meat Postslaughter (Rigor Mortis) — 36
        - 2.2.1 Anaerobic Glycolysis and the Formation of Lactic Acid — 36
        - 2.2.2 ATP Levels Postslaughter and Rigor Mortis — 38

3. **Definitions**
    - 3.1 PSE/RSE/DFD and "Normal" Meat — 39
        - 3.1.1 PSE/RSE/DFD Meat — 40
    - 3.2 Mechanically Separated Meat — 43
    - 3.3 Cold Shortening — 44

3.4	Freezing and Tempering of Meat	46
	3.4.1 Tempering of Meat	48
3.5	Freezer Burning	49
3.6	pH Value	49
3.7	Aw Value	51
3.8	Eh Value (Redox Potential)	52
3.9	Condensation Water	53
3.10	Maillard Reaction	54

Part II
Additives

4. Additives

4.1	Phosphates	59
	4.1.1 Production and Properties of Phosphates	60
	4.1.2 Properties of Different Phosphates	60
	4.1.3 P_2O_5 Content of Phosphates	62
4.2	Salt—Sodium Chloride	62
4.3	Potassium Chloride and Sodium-Reduced Sea Salts	64
	4.3.1 Sodium-Reduced Sea Salt	65
4.4	Caseinate and Whey Protein	65
	4.4.1 Whey Protein	65
4.5	Soy Protein	66
	4.5.1 Gelation of Soy	67
4.6	Pork Rind Powder	68
4.7	Sugars	68
4.8	Antimold Materials	70
4.9	Monosodium Glutamate	71
4.10	Ribonucleotide and Other Flavor Enhancers	72
4.11	Water/Ice	72
4.12	Spices and Spice Extracts	74
4.13	Hydrolyzed Vegetable Protein	78
4.14	Antioxidants	78
4.15	Natural Smoke	80
4.16	Liquid Smoke	84
4.17	Colors	84
4.18	Chemical Nonbacterial Acidulants	86
	4.18.1 Glucono-δ-lactone	86
	4.18.2 Encapsulated Citric Acid	87
4.19	Fibers	87

5. Color in Cured Meat Products and Fresh Meat

5.1	Retention of Color in Fresh Meat and Uncured Meat Products	89
5.2	Nitrite and Nitrate	93
	5.2.1 Natural Nitrite	96
5.3	Mechanism of Color Development in Cured Meat Products	96

		Contents	vii

	5.4	Color Enhancers	99
	5.5	Measuring Color: L*-a*-b* System	101

6. Casings
	6.1	Natural Casings	103
		6.1.1 Collagen Casings	106
	6.2	Artificial Casings	107

Part III
Production Technology

7. Fermented Salami: Non–Heat Treated
	7.1	Selection of Raw Materials	112
		7.1.1 Meat	112
		7.1.2 Fat	117
	7.2	Selection of Additives	119
		7.2.1 Selection of Acidification Additives	123
		7.2.2 Starter Cultures	125
		7.2.3 Selected Wanted Mold and Yeast	130
	7.3	Manufacturing Technology	133
		7.3.1 Salami Made in a Bowl-Cutter	133
		7.3.2 Salami Made With Mincer-Mixing System	136
		7.3.3 Salami Made With Mincer-Filling Head	137
		7.3.4 Filling	138
		7.3.5 Fermentation	141
		7.3.6 Drying of Salami	162
		7.3.7 Slicing	171
		7.3.8 Packaging/Storage	171
	7.4	Summary of Critical Production Points	174
		7.4.1 Microbiology in Salami	175

8. Typical Fermented Non-Heat Treated Salami Products Made Around the World
	8.1	Hungarian Salami	177
	8.2	Kantwurst (Austria)	178
	8.3	German Salami	179
	8.4	Lup Cheong (China)	179
	8.5	Cacciatore (Italy)	180
	8.6	Sopressa (Italy)	181
	8.7	Milano Salami (Italy)	181
	8.8	Sucuk (Turkey)	182
	8.9	Chorizo (Spain)	182
	8.10	Fuet (Spain)	183
	8.11	Pepperoni (USA)	183
	8.12	Pizza Salami (Australia)	183
	8.13	Beef Salami (Malaysia)	184

9. Fermented Salami: Semicooked and Fully Cooked

9.1 Manufacturing Technology   185

10. Typical Fermented Semicooked and Fully Cooked Salami Products Made Around the World

10.1 Summer Sausage (USA)   187
10.2 Pizza Salami (Australia)   187
10.3 Poultry Salami (Asia)   188

11. Nonfermented Salami: Fully Cooked

11.1 Selection and Preparation of Raw Materials   189
11.2 Selection of Additives   190
11.3 Manufacturing Technology   191
    11.3.1 Precuring of Meat and Fat Materials   191
    11.3.2 Manufacture of BE   192
    11.3.3 Finely Cut Cooked Salami   192
    11.3.4 Coarse Minced-Mixed Cooked Salami   194
    11.3.5 Filling   195
    11.3.6 Drying, Smoking, and Cooking   196
    11.3.7 Packaging and Storage   197
    11.3.8 Summary of Critical Production Points   198

12. Typical Nonfermented Salami Products Made Around the World

12.1 Polish Salami (Austria)   199
12.2 Cheese Salami (Austria)   199
12.3 Vienna Salami (Austria)   200
12.4 Salami Florentine (Austria)   200
12.5 Cabanossi (Austria)   200
12.6 Strasburg Salami (Australia)   201
12.7 Chicken Salami (Asia)   201

Index   203

Preface

Salami, or salami-type products, have been produced for a very long time all over the world. *Salumi* is the Italian word for "salted and cured meats" (which includes salami as well as products such as prosciutto), while *salami* refers to dry-cured sausages. *Salame* is the word for a single salami. Despite the rich history and versatility of salami, it is virtually impossible to find a book covering the science of salami in combination with clearly outlined processing steps in order to achieve safe and great-tasting end products. Salami, and especially fermented and dried salami, still awakens some agree of uneasiness in people because such products are produced from raw meat and the end product is consumed while still raw. Also, microbiology plays a vital role in the production of fermented and dried salami, which complicates matters even more.

The purpose of this book is to give clear and helpful guidelines to professionals within the meat-processing industry, such as technical, production, operations, process improvement, quality control, and research and development managers. Undergraduate as well as postgraduate students and academicians will find this book an invaluable tool for their studies and lectures.

Having worked all over the world in the meat-processing industry, conducting seminars for customers as well as lecturing at universities, I was frequently asked about the availability of an all-in-one book about salami. There are very few books available on the topic of salami, and most are written for hobbyists, mostly describing recipes of products but not focusing on science and professional processing technology. This book fills a gap because it combines a scientific and yet still hands-on approach allowing for the safe and efficient production of all salami-type products. Microbiology related to salami is also discussed as it plays a vital role in the production of salami.

Acknowledgments

This book summarizes both my practical and my theoretical knowledge gained from working within the meat industry in several countries during the past 30 years. The practical operations knowledge, including process optimization, is based on having occupied roles as such production and factory manager in three countries, as well as having had my own consultancy business for process optimization in the meat and food industry. My theoretical knowledge gained over the years is the result of both having been taught by and having worked with extremely knowledgeable people during my study of the Master Butcher Diploma and especially during my study of Meat Technology in Kulmbach (Germany). Therefore, I want express deep gratitude to those who taught me over all those years, namely Titus Kaibic, Dipl.-Ing. Thomas Eberle, Dr. Gerhard Hartmann, Dr. Fredi Schwaegele, Dipl.-Ing. Hans-Georg Hechelmann, Professor Dr. Lothar Leistner, Dr. Hermann Hecht, Dr. Wolfgang Schneider, Dr. Ulrike Fischer, Dr. Andrea Maurer, and Dr. Peter Braun. I also want to thank all those people whom I no longer even remember who gave me ideas and help over the years. However, the biggest "thank you" must go to my patient wife, Tracy, and our two sons, Jack and Lewis, for their wonderful support while I was researching, writing, and editing this book.

During the course of my studies, several books were of great help, namely:

Breuer, H. *dtv-Atlas zur Chemie II: Organische Chemie und Kunststoffe*, 5th edition. Deutscher Taschen Verlag, Muenchen, 1992.

Kulmbacher Reihe, Band 2, *Beitraegezur Chemie und Physik des Fleisches*. Foerderungsgesellschaft der Bundesanstalt fuer Fleischforschung, Kulmbach, 1981 (articles by K. Hofmann, K. Potthat, R. Hamm, K. Fischer, K.O. Honikel, and L. Toth).

Kulmbacher Reihe, Band 5, *Mikrobiologie und Qaulitaet von Rohwurst und Rohschinken*. Foerderungsgesellschaft der Bundesanstaltfuer Fleischforschung, Kulmbach, 1985 (articles by W. Roedel, F.-J. Lueke, and H. Hechelmann).

Kulmbacher Reihe, Band 6, *Chemisch-physikalische Merkmale der Fleischqualitaet*. Foerderungsgesellschaft der Bundesanstalt fuer Fleischforschung, Kulmbach, 1986 (article by K.O. Honikel, H. Hecht, and K. Potthast).

Kulmbacher Reihe, Band 10, *Sichere Produktebei Fleisch und Fleischerzeugnissen*. Foerderungsgesellschaft der Bundesanstalt fuer Fleischforschung, Kulmbach, 1990 (article by L. Leistner, H. Hechelmann, R. Kasprowiak, R. Geisen, F.-K. Lueke, and L. Kroeckel).

Mueller, G. *Grundlagen der Lebensmittelmikrobiologie*. Fachbuchverlag (on behalf of Dr. Dietrich Steikopf Publishing), Leipzig, 1986.
Praendl, O., Fischer, A., Schmidhofer, T., and Sirell, H.-J. *Fleisch*. Eugen Ullmer, Stuttgart, 1988.

Introduction

Humans have eaten meat products for centuries, and salami-type sausages are the oldest form of "sausage." Drying of meat and meat products as a form of preservation dates back to the ancient Chinese, and countless highly valuable vitamins, minerals, and trace products can be found in meat as well as meat products. Despite ongoing discussions about health concerns when consuming meat and meat products, it can be said that these types of food are still part of a balanced and healthy diet today. There are also those, though, who opt not to eat meat for a variety of reasons.

Disclaimer

All information in regards to practical knowledge gained by the author while working in factories as well as theoretical knowledge gained during his studies should not be used as the basis for any legal claims. Hence, all information stated is not intended to credit, or discredit, any manufacturer of equipment or suppliers of meat or additives and is purely based on the opinion of the author.

Part I

Meat and Fat

Chapter 1

Meat and Fat

1.1 INTRODUCTION

Most countries interpret the presence of "pork meat" in their respective food standards as meaning "muscle meat including fat and skin" rather than as lean muscle tissue only. This fact can be confusing to the ordinary consumer as most understand lean muscle tissue to be "meat" and do not know that fat and skin are classified as "meat" as well. When other types of meat are described, though, the term meat does not include fat and skin. The amount of lean meat obtained from a carcass is around 35% in cattle, around 45% in pigs, around 38% in veal, and around 35% in lamb. Fat is also part of a balanced human diet, and the presence of fat in meat and meat products has both technological and organoleptic purposes. The relationship between fat consumption and weight gain, though, is currently a topic of interest, as excessive consumption of fat may be a cause of the increased levels of obesity worldwide.

The quality of meat and meat products is also a topic of frequent discussion. There is currently no consensus on what the term "quality" really stands for given that "quality" is generally seen as a combination of two major aspects. "Total quality" of meat and meat products includes, on the one hand, characteristics that can be measured, such as microbiological status, tenderness, color, juiciness, shelf life, pH value, pesticide levels, etc. On the other hand, total quality also includes an aspect that is less easy to measure: the consumer's personal perception of the value of meat and meat products. This perception is different for every individual human being, as external factors such as television advertisements have an influence on this aspect of total quality. The term "quality," from the consumer's point of view, could be simply said to mean whether the consumer thinks a product is good value for money, and this judgment will vary from person to person and from product to product.

The study of meat technology evolves around the five major building blocks used to make a meat product: raw materials, additives, the manufacturing technologies applied, food safety, and commercial interests, including all possible interactions between the five. Manufacturing technology combines raw materials and additives with each other to obtain a product of the desired quality within a certain economic framework possibly providing profit to the manufacturer of the product (see Fig. 1.1).

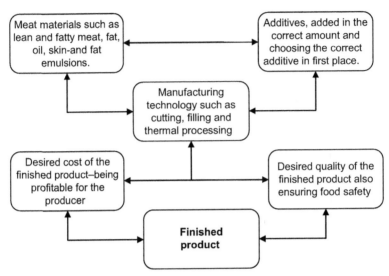

FIGURE 1.1 Overview on meat technology.

1.2 AMINO ACIDS

Amino acids are the building blocks of proteins. Even though about 190 amino acids are known today, only 20 different amino acids are required by humans to synthesize all necessary proteins. All of these 20 amino acids are α-amino acids given that both functional groups, the "acid" carboxyl group (—COOH), and the "alkaline" amino group (—NH$_2$), are attached to the same carbon atom, the alpha (α)-carbon atom or Cα. This alpha-carbon atom is also referred to as the "chiral center" and glycine, the simplest amino acid, is the only nonchiral amino acid. The rest of the molecule is in most cases the primary portion of the amino acid and determines the identity of the amino acid itself as well as whether the amino acid is polar or nonpolar.

As stated, almost all α-amino acids are chiral, meaning that two arrangements of the same molecule are nonidentical mirror images.

Chiral amino acids exist in two configurations known as L- or D-stereoisomers, which correspond to "left-handed" (L) or "right-handed" (D) three-dimensional shapes. D originates from the Latin word *dexter* and the NH$_2$ group is on the right hand side of the molecule, while L originates from *laevus*, meaning "left." All amino acids found in proteins are L-isomers except for glycine, the simplest amino acid, which is not chiral. Depending on the side chains within the amino acid, neutral, acid, or alkaline amino acids are formed (see Fig. 1.2).

Amino acids exhibit side groups, which can be made of a hydrogen atom or other ring-structured molecules. In turn, those side groups can show different groups such as hydroxyl groups (—OH) and, in conjunction with the carboxyl

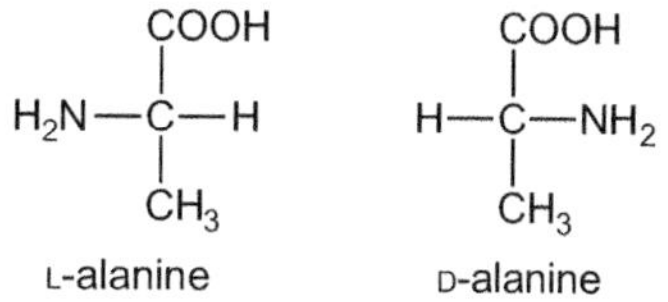

$$H_2N-\underset{\underset{CH_3}{|}}{\overset{\overset{COOH}{|}}{C}}-H \qquad H-\underset{\underset{CH_3}{|}}{\overset{\overset{COOH}{|}}{C}}-NH_2$$

L-alanine   D-alanine

FIGURE 1.2 L- or D-form of the amino acid alanine.

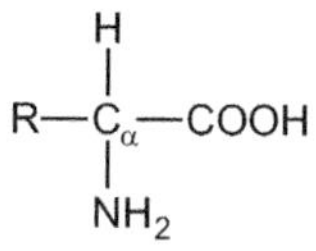

$$R-\underset{\underset{NH_2}{|}}{\overset{\overset{H}{|}}{C_\alpha}}-COOH$$

FIGURE 1.3 Typical configuration of an amino acid; R represents the "rest" of the molecule.

and amino group of the main structure of the amino acid, these affect the structure of a protein (see Fig. 1.3).

Eight of those 20 amino acids are "essential" and have to be supplied to the human body by consuming food, containing those essential amino acids.

The body cannot synthesize those eight essential amino acids, and if they are not provided to the human body via the intake of food, illness and death may be the consequence. The remaining 12 amino acids can be synthesized by the human body, as long as food consumed provides all elements needed to synthesize those amino acids. Protein-containing food is broken down via digestion into individual amino acids, from which the required body proteins are synthesized.

The eight essential amino acids are:

- Isoleucine
- Leucine
- Lysine
- Methionine
- Threonine
- Valine
- Tryptophane
- Phenylalanine

All other 12 amino acids can be synthesized by the human body itself using nitrogen, which is supplied by consuming food containing nitrogen.

- Alanine
- Arginine
- Aspartic acid
- Proline
- Tyrosine
- Serine
- Asparagine
- Cysteine
- Glutamic acid
- Histidine
- Glutamine
- Glycine

The nutritional value of food is determined by the presence of essential amino acids at their lowest relative concentration. A food might contain seven of the eight essential amino acids at a high concentration but one at only a very low level, and it is the one present at a low level that determines the nutritional

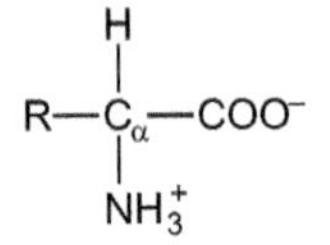

$$R-\underset{\underset{NH_3^+}{|}}{\overset{\overset{H}{|}}{C_\alpha}}-COO^-$$

FIGURE 1.4 Configuration of a zwitterion.

value of the food. This is based on the fact that if only this particular type of food were consumed to provide essential amino acids, the one present at the low concentration would be always "missing" and illness would be the result given that the body cannot synthesize this one essential amino acid.

Amino acids are "weak acids" present in a solution as zwitterions at a pH value of 5.4–5.7. In such a situation, the COOH group is present as a negatively charged COO^- ion and can take up a hydrogen (H^+) ion, while the NH_2 group is present as a positively charged NH_3^+ ion, which can give away, or donate, one H^+ ion (see Fig. 1.4).

Amino acids can act as an "acid" or as an "alkaline" depending on their pH environment. At low pH values, or in a sour environment, the negatively charged carboxyl group (COO^-) can take up a H^+ ion and act as an "alkaline." The gain of a hydrogen ion neutralizes the COOH group, and the entire amino acid becomes positively charged due to the excess H^+ ion on the NH_3^+ group. During high pH values or in an alkaline environment, the positively charged NH_3^+ group of an amino acid releases a H^+ ion and acts therefore as an "acid." The NH_2 group is neutralized and the entire amino acid becomes negatively charged due to the COO^- group still present within the amino acids. Amino acids can donate or absorb H^+ ions without changing their pH value, which explains the "buffer capacity" of amino acids and, subsequently, proteins. The buffer capacity depends on the concentration of ions donated or absorbed, and once the buffer capacity is exceeded, the pH value of the protein will change.

1.3 PROTEINS

Proteins belong to the class of organic compounds called polyamides and are condensation polymers of amino acids. Proteins consist of carbon (C), oxygen (O), hydrogen (H), and nitrogen (N), and some contain sulfur, phosphorous, and iron. Expressed in percentages, proteins contain around 52% carbon, 19% oxygen, 16% nitrogen, 6% hydrogen, and some sulfur. Proteins are macromolecules and are formed by amino acids joined together through the reaction of an acid carboxyl group (—COOH) on one amino acid and an alkaline amino group (—NH_2) on the other amino acid. Water is eliminated during this condensation reaction and an input of energy is required for the reaction to occur in first place. More specifically, the eliminated water (H_2O) is formed out of an OH part from the carboxyl group and an atom of hydrogen (H) from the amino group. The

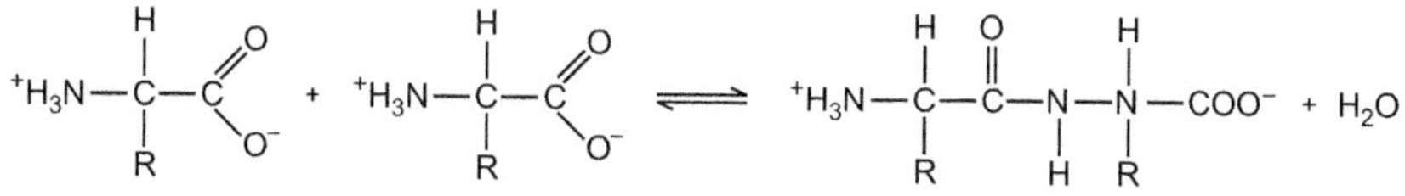

$$^+H_3N-\underset{R}{\underset{|}{\overset{H}{\overset{|}{C}}}}-C\overset{O}{\underset{O^-}{\diagup}} \; + \; ^+H_3N-\underset{R}{\underset{|}{\overset{H}{\overset{|}{C}}}}-C\overset{O}{\underset{O^-}{\diagup}} \; \rightleftharpoons \; ^+H_3N-\underset{R}{\underset{|}{\overset{H}{\overset{|}{C}}}}-\overset{O}{\overset{\|}{C}}-\underset{H}{\underset{|}{N}}-\underset{R}{\underset{|}{N}}-COO^- \; + \; H_2O$$

FIGURE 1.5 Peptide bond formed by the joining of two amino acids.

links formed between two amino acids are known as peptide links, or peptide bonds, and the link shows a characteristic CO—NH bridge. A peptide consists out of two amino acids bound together while oligopeptides show up to 10 amino acids within their structure (see Fig. 1.5).

Peptides containing between 50 and 80 amino acids are called polypeptides and peptides, and those made of more than 80 amino acids are proteins. Each protein has its own molecular weight (M_r), which is measured relative to the mass of 1 atom of ^{12}C. The molecular mass is expressed in Daltons (Da) or kilo-Daltons (kDa), and 1 Da is 1/12 the mass of a atom of ^{12}C (1.6×10^{-24} g). The molar mass, on the other hand, is the mass of 1 mol expressed in grams, and 1 mol of a substance is made of 6×10^{23} atoms of the same kind. Proteins vary in solubility and basically do not have a color or taste. The digestibility of meat protein is around 95, the same as milk and egg, compared with around 85 from plant proteins. The biological value of meat protein is 0.76 compared to 1.0, the biological value of human milk. In proteins, bound together by peptide bonds, the hydrophilic (water-loving) groups are facing in, while the lipophilic (fat-loving or hydrophobic) groups are facing out. This leads to different structures of proteins.

Proteins can show four different structures, as shown next.

1. Primary structure
   The primary structure of a protein is its peptide "backbone" or polypeptide chain, formed through peptide bonds. The polypeptide chain consists of a unique sequence of amino acids, which is genetically determined. The polypeptide chain is linear, not branched and no other bonds are involved or forces implied in this structure.
2. Secondary structure
   The secondary structure is a regular repeating folding pattern, either an α-helix or β-pleated sheet, stabilized by hydrogen bonds between amide groups of the peptide bonds, which are present along the chain of amino acids and other carbonyls. Hydrogen bonds show the characteristic H—H bridge. An α-helix is right-handed, and the β-pleated sheet structure is formed by the assembly of extended polypeptide chains lying side by side. The formation of either an α-helix or a β-pleated sheet depends on the sequence of amino acids in the primary structure. Overall, the secondary structure is a localized, repetitious folding or twisting of the polypeptide chain in the primary structure.

3. Tertiary Structure
   Secondary structures unfold into three dimensions, giving rise to tertiary structures. Tertiary structures are supported by binding forces such as ionic bonds, hydrogen bonds, disulfide bridges, Van der Waals forces, and hydrophobic interactions. In tertiary structures, the folding is not predictable or repetitive, as it is the case within the secondary structure.
4. Quaternary Structure
   A quaternary structure is obtained when two or more individual polypeptide chains function as a single unit. The interactions between polypeptides create an oligomeric structure stabilized by noncovalent bonds. Loose hydrogen and sulfide bonds are present as well and the overall shape can be fibrous (threadlike) or globular. A well-known representative of a quaternary structure is hemoglobin, where the globin is surrounded by four heme molecules.

Complex structures such as the secondary, tertiary, and quaternary structures are easily changeable. Changes in structure are known as denaturation, which can take place by the impact of temperature (cooking), pH values (acidification), high concentrations of salt (salting), or low levels of water activity. As a consequence of denaturation, the protein changes its configuration from a highly organized and native structure into a less-organized (denatured) and non-native structure. Due to this change in three-dimensional configuration, the protein loses its native form and is not "functional" anymore. The process of denaturation is generally irreversible and the tertiary and quaternary structures are primarily affected. Often, though, especially in irreversible denaturation, the secondary structure is affected as well, but the primary structure is not affected during the process of denaturation.

Analyzing meat products with regard to their protein content takes place in most countries based on the Kjeldahl method. This method is based on "finding" the total nitrogen within the meat product given the fact that proteins contain a certain percentage of nitrogen (16%). Once the "total nitrogen" is determined, the figure is multiplied by a factor 6.25, which results in the amount of protein present within meat or the meat product. The factor 6.25 originates from dividing 100 (total protein) by 16 (16% of nitrogen), which gives, as said, the factor 6.25 (see Fig. 1.6).

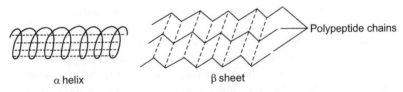

FIGURE 1.6   The α-helix and β-pleated sheet secondary structure.

1.4 COLLAGEN

Collagen is a substantial part of connective tissue and is found in ligaments, tendons, skin, and many other types of tissues serving mechanical and structural functions. Collagen accounts for almost one-third of the total protein and is made out of several different proteins. One of the major components is the amino acid hydroxiproline, and this particular amino acid is present within collagen in a concentration of 12.5%. The amino acids proline and glycine are present at around 45–50% within collagen.

Collagen is stabilized against mechanical forces by cross-linking, which is induced by enzymes. Cross-links contribute to the enormous strength of collagen fibers, and the enzyme responsible for the formation of those cross-links is lysyl oxidase. Another form of stabilization within collagen comes from the presence of hydrogen as well as other covalent bonds. The strength of collagen increases with age of the animal, and the solubility of collagen, in old animals, is reduced as a higher number of cross-links are formed within the collagen molecule at increased age. An analysis toward the content of collagen in a meat product is based in the first place on determining the amount of hydroxiproline present. The figure obtained of hydroxiproline is multiplied by the factor of 8, which is based on the above mentioned fact, that collagen contains 12.5% hydroxiproline. Dividing 100 by 12.5, the factor 8 is obtained.

The building blocks for collagen are units of tropocollagen, which is a right-handed triple helix made out of three intertwined proteins, and numerous molecules of tropocollagen align themselves next to each other to form collagen. During this process, some amino acids are removed from individual tropocollagen molecules, which allows them to align next to each other to form collagen. Cross-links on the side chain of the amino acids lysine and histidine, as well as hydrogen bonds, stabilize those helices of tropocollagen within collagen. Such cross-links are unusual in proteins and only occur in collagen and elastin. A triple helix of tropocollagen is, further, made out of three individual helices called procollagen. Each individual strand of procollagen exhibits a left-handed helix within itself. Thus, tropocollagen could be called a "coiled coil" given that left-handed individual helices of procollagen form the right-handed triple helix of tropocollagen, and such double-coiling is ultimately responsible for the enormous strength to collagen. Every third amino acid within collagen is glycine, the smallest of all amino acids, and its presence helps to stabilize the triple helix. Collagen is insoluble in water and is not soluble by the impact of salt and/or phosphates, while procollagen is water soluble. The triple helix swells if exposed into a sour media for a prolonged period of time and "absorbs," or holds, water during this process of swelling.

Collagen, exposed to moist heat for a prolonged period of time, turns into gelatin, which forms a gel on cooling. Prolonged periods of heat treatment turn gelatin again into individual strands of procollagen, which, contrary to gelatin, does not form a gel on cooling. The "solubility" of collagen during heat

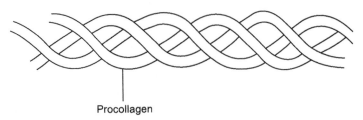

Procollagen

FIGURE 1.7 Triple helix of tropocollagen made out of individual strands of procollagen.

treatment depends greatly on the number of cross-links present within triple helices (hydrogen and covalent bonds), and increasing age of an animal leads to a higher number of cross-links, which in turn reduces solubility. As a result, hydrogen and as covalent bonds within collagen, which stabilize the molecule in first place, reduces solubility at the same time. Collagen becomes "tender" in raw dried products, such as prosciutto, by the impact of the enzyme collagenase, which is able to soften collagen over a prolonged period of time during ripening and drying of such products (see Fig. 1.7).

Elastin is another component present within connective tissue at around 4% of the amount of collagen and around 0.8% from the total meat protein. The amino acid hydroxiproline is found within elastin at 1% and elastin is yellowish in color, practically insoluble in water and salt and is resistant toward diluted acids.

1.5 MUSCLE PHYSIOLOGY

A single muscle is covered by a thin layer of connective tissue called the epimysium, which is the extension of the tendon. Muscle is divided into muscle fiber bundles, and another thin layer of connective tissue, called perimysium, covers each fiber bundle. In turn, each fiber bundle is made of individual muscle fibers, which are covered by a membrane of connective tissue known as endomysium. Under the endomysium is another layer known as sarcolemma, which is of net-like structure and is directly connected to the filaments actin and myosin, the major components of a muscle fiber. A liquid, called sarcoplasm (cytoplasm), is the intracellular substance in a muscle fiber and consists of around 80% water as well as proteins, enzymes, lipids, carbohydrates, inorganic salts, as well as metabolic byproducts (see Fig. 1.8).

Lean muscle tissue contains 70–75% water, 22% protein, around 1–3% intramuscular fat, and around 2% of other components such as phosphates, ash, and minerals. Lean meat also contains around 45–50 mg of cholesterol per 100 g of muscle tissue. The 22% of total protein can be divided into around 13–14% myofibrillar protein (salt soluble), 7% sarcoplasmic proteins (water soluble or soluble at very low salt concentrations), and around 2% structural proteins

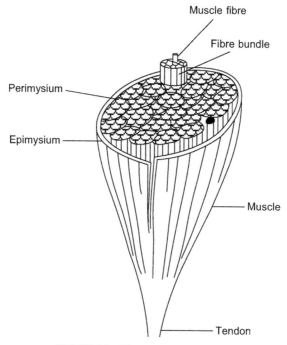

FIGURE 1.8 The structure of a muscle.

such as connective tissue (insoluble in salt and water). Expressed in percent, the protein in lean muscle tissue consists of 65% myofribrillar protein (myosin around 42% and actin, troponin, and tropomyosin together around 23%), around 25% sarcoplasmic protein, and around 10% connective tissue. As seen, actin and myosin account only for around 6.5% of the total muscle weight, and only around 65 g/kg of meat is soluble protein coming from actin and myosin. Actin and myosin are also known as the myofilaments and are responsible for muscle contraction and relaxation.

Albumins and globulins are the main sarcoplasmic proteins, and around 90 different proteins belong to the group of sarcoplasmic proteins. Albumins are fully soluble in water, while globulins are soluble in weak salt solutions only but insoluble in water. Myoglobin (color of meat) and hemoglobin (color of blood) are most important types of globulins. Sarcoplasmic proteins are responsible for the metabolism in an animal cell. The main representative in the group of connective tissue is by far collagen (40–60%), and some tropocollagen, as well as elastin (around 10%), is present as well. Hence, a major part of connective tissue, around 30%, is made of other insoluble proteins. Myofibrillar proteins denature at around 67–72°C, while sarcoplasmic proteins denature generally at around 62–70°C. Some sarcoplasmic proteins denature at temperatures as low as 50°C.

Connective tissue shrinks by 60–65°C, and continuous moist heat treatment up to around 90–95°C turns collagen into gelatin. As said earlier, prolonged heat treatment turns gelatin into individual strands of procollagen.

Water within muscle tissue is bound more or less firmly and in different ways. Protein-bound water within meat, around 4–6% of water within muscle tissue overall, is bound so firmly to protein that even by a temperature of around −45°C, protein-bound water is still not frozen. Around 55–60% of water, present between the myofibrils, is bound in relation to the pH value of meat. Such fibril-bound water is known as immobilized (or not freely available) water but is not bound as firm as protein-bound water. Water present in the sarcoplasm, around 20–25%, is freely available and known as "free water." Last, extracellular water accounts for around 8–14% of the total water and is held outside cellular membranes in capillaries.

The contractile unit of a muscle fiber is a sarcomere and, for example, the human biceps exhibits around 10 billion sarcomeres (see Fig. 1.9). A sarcomere, which is around 2 μm long, lies between two Z-lines. Actin is connected to the Z-line and comes to an end there. Myosin, on the other hand, is connected to the M-line. The I-band is the zone where no myosin overlaps with actin, and the H-zone is the space where no actin overlaps with myosin. The A-band represents myosin. The special arrangement of actin and myosin give the fiber a striated appearance under the microscope, and actin and myosin are arranged in a structured hexagonal pattern.

Myosin, the "thick" filament, is made of around 280 molecules of individual units of the protein myosin, which exhibits a long tail as well as two pear-shaped heads. The molecular weight of myosin is around 490,000 Da, and the tail of the molecule is called light myosin, while the head is known as the heavy myosin. The isoelectric point (IEP) of myosin, where negative and positive charges are of the same number within the protein, is by a pH value of 5.0. Myosin is

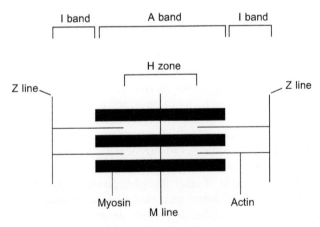

FIGURE 1.9 Sarcomere.

soluble in salt concentrations ranging from 1% to 5%. Above a salt concentration of 5%, myosin is denatured due to the high level of salt and is not native (functional) anymore. The tail of the myosin molecule is rich in acid as well as alkaline amino acids, and the COO^- and NH_3^+ groups are attracted to each other inside the tail. Those charged ions hold the thick filament together and are responsible for the limited solubility of myosin in water. Myosin also carries the enzyme myosin ATPase, which plays a major in role in the movement of muscle fibers during contraction and relaxation of a muscle (see Fig. 1.10).

Actin, the "thin" filament, consists of monomeric or globular G-actin, which forms the filamentous, or fibrillar, F-actin, as well as troponin, tropomyosin, and actinin. The molecular weight of G-actin is around 40,000 Da, and it is made of around 350 amino acids. Around 360 molecules of G-actin are required to polymerize and to form F-actin, which are two strings of actin globules wound around each other in a double helix like a pearl chain. Such a pearl chain of F-actin is curled around the string-like tropomyosin, which is made of two curled strings on its own. After every seventh actin molecule, a troponin complex made of troponin I, C, and T is attached to actin, and such troponin exhibits a high affinity toward Ca^{2+} ions. Troponin and tropomyosin have a controlling, or regulating, impact on the contraction and relaxation of the muscle fibers but are not directly involved in contraction and relaxation as such. Denaturation of those contractile fibers myosin and actin starts by 55–60°C (see Fig. 1.11).

The amount of water bound within muscle tissue depends to a large degree on the space available between the fibers actin and myosin, and the pH value plays a vital role. Generally, pH values above and below the IEP result in enhanced water-holding capacity (WHC), but within muscle tissue, the levels above the IEP are of importance. A decline in pH value and nearing of the IEP in meat result in less WHC. At the IEP of the actin-myosin complex, by a pH value of 5.2, most COOH groups are present as COO^- anions and most of NH_2 groups are present as NH_3^+ cations. Those positive and negative ions attract

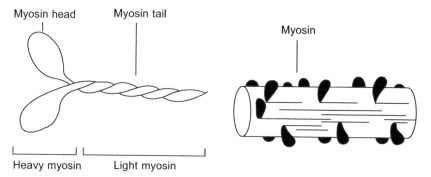

FIGURE 1.10 Myosin as a single unit and also in its "thick" state, made of countless individual units.

each other, and the protein molecule is tightly bound together. At this point, the protein molecule shows a net charge of zero as positive and negative charges present within the protein molecule are of the same number. As a result, only a tiny amount of water can be bound within the proteins. An increase in number of charges, positive or negative, within the protein molecule increases the WHC given that the protein is not as tightly bound together as it is the case at the IEP. When negative charges outnumber the positive charges, the pH value of the protein is above the IEP and such a condition results in increased WHC. The fibers (filaments) are repulsed and the space, or gap, between actin and myosin is enlarged, which allows more water to be incorporated. Within fresh meat, a very similar effect can be seen at a pH value below the IEP when positive charges outnumber the negative charges (see Fig. 1.12).

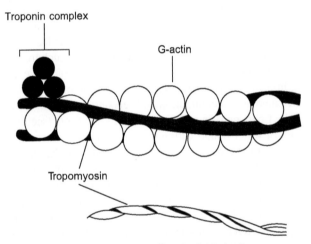

FIGURE 1.11 Actin made of individual units.

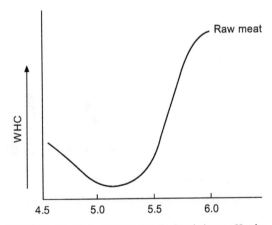

FIGURE 1.12 Water-holding capacity in relation to pH value.

In the living animal, some of the NH_3^+ groups are present in muscle tissue as neutral NH_2 groups and binding forces between actin and myosin are weaker. When the binding forces are weaker, enhanced levels of water can be immobilized within the fiber structure, and such immobilized water is not freely available but firmly bound. Hence, to the increased repulsion between actin and myosin, the capillary effect is enhanced as well, which supports uptake of water once again. The capillary effect can be compared with a slight "sucking" effect and is seen if a straw is placed in a glass of water as the level of water within the straw is above the level of water in the glass. The positive impact of the capillary effect regarding WHC of muscle tissue comes to an end at a pH value around 6.4, as the individual fibers are so far apart from each other at this point that the "sucking effect" no longer takes place. As a result, an elevated pH value (increase in negative charges) in muscle tissue creates larger gaps between actin and myosin due to increased repulsion forces, combined with the capillary effect, and more water can be immobilized (see Fig. 1.13).

The introduction of salt into muscle tissue changes the number of charges on the muscle fibers, which in turn causes the swelling of the fiber structure and WHC is enhanced as a consequence. Positively charged sodium ions (Na^+) are bound fairly weakly by negative charges within the protein molecule. On the other hand, the negative chloride ion (Cl^-) is bound strongly within the molecule and becomes neutralized or gets basically taken out of the equation. The "light" binding of the sodium ion causes a slight movement of the IEP within the muscle meat toward a lower pH value, around 5.0, and at the same time creates a bigger space between the fibers at a certain pH value present in meat (see Fig. 1.14).

1.6 FLAVOR OF MEAT

Generally, raw meat itself does not demonstrate a great deal of flavor, and flavor intensity increases with the age of an animal regardless of the type of animal. For example, beef from an animal that is only 1 year old does not exhibit as much

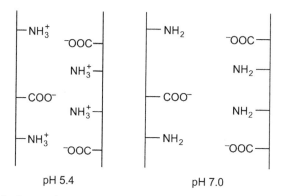

FIGURE 1.13 Impact of the pH value regarding repulsion forces between actin and myosin.

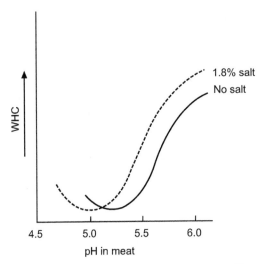

FIGURE 1.14 Increased WHC due to the addition of salt by moving the IEP toward the acid area.

beef flavor as beef originating from a 2-year-old cattle. The same principle of age regarding flavor applies to pigs and poultry. Generally, meat from an older animal of the same species exhibits stronger flavor compared to a young animal.

Meat contains around 1.3 g of sodium chloride per kilogram as well as 3 g of potassium, and both contribute to the slightly "salty" and "bitter" tastes of raw meat. There is little sugar in meat except in meat originating from horse (around 1%) or deer such as venison. The "sour" component of meat is primarily based on lactic acid obtained during post mortem glycolysis. Another contribution to flavor comes from the degradation of adenosine diphosphate (ADP) finally into inosine monophosphate (IMP), which is ultimately converted into inosine and hypoxanthine. Both substances add some degree of "bitter" flavor to meat. This aspect of added bitterness is present in a stronger form within DFD meat. Given that less lactic acid is produced in DFD beef meat (see Section 3.1 under Chapter 3: Definitions) during rigor mortis, the overall flavor is not as strong as in "normal" beef meat. Hence, IMP, originating out of ADP during post mortem glycolysis and present in meat, is converted in DFD meat to a larger extend into hypoxanthine, a slightly bitter-tasting substance. The flavor of raw meat is also influenced to some degree from sulfuric components present in meat. Other flavor components in meat are amino acids, peptides, and carbonic acids. During aging (ripening, maturing) of meat, lipolysis (breakdown of fats) and proteolysis (breakdown of proteins) cause proteins and fat to become a much more significant part of meat flavor. Pork meat also contains several different lactones, which contribute to the typical pork flavor.

Once meat is cooked, the flavor intensifies with substances such as alcohols, aldehydes, ketones, and lactones generating flavors. Pork fat produces, on heating, two saturated as well as two unsaturated aldehydes, which are not present in heated beef, chicken, or lamb fat. Heat treatment by the impact of high temperatures such as grilling creates ring-shaped molecules, responsible for the typical grill flavor contrary to saturated ketones, aldehyde, and acids produced during low-temperature heat treatment obtained by processes such as steam cooking and simmering. Through the impact of heat, the Maillard reaction (see Section 3.10 under Chapter 3: Definitions) takes place in meat, and the presence of the sulfur-containing amino acids cysteine and methionine leads to very flavorful components. Hence, carbonyl components such as oxopropanol and methylfurfural, obtained from the Maillard reaction, contribute greatly to flavor of cooked meat as well.

Boneless cuts of meat are commonly packed under vacuum for easier and prolonged storage as well as hygienic handling. Once removed from the packaging, such meat has to be exposed for around 10 minutes to fresh air to regain its original flavor. Within vacuum packaging, an atypical flavor is created as a result of the vacuum applied.

PSE and DFD meat (see Section 3.1 under Chapter 3: Definitions) exhibit a reduced amount of glycogen, and by preparing food with such meat, the Maillard reaction takes place in a weaker form, creating less flavor as a result. Boar meat contains the sex hormone androsterone, which creates a "flavor" widely disliked by people, especially by females. Chicken meat receives its flavor primarily from unsaturated aldehydes, which originate from linoleic acid in the first place. Occasionally, a slight "fishy" flavor can be observed in chicken and turkey, and the cause for that seems to be feed containing high levels of unsaturated fats.

1.7 PRINCIPLES OF MUSCLE CONTRACTION AND RELAXATION

In a relaxed muscle, the level of Ca^{2+} ions is very low and the troponin present does not allow the myosin head to bind to the actin. As a result, the muscle stays relaxed. The contraction of a muscle is heavily regulated by Ca^{2+} ions, released from the sarcoplasmic reticulum (SR), based on a nervous impulse. As a result of such as nervous impulse, the concentration of Ca^{2+} ions increases in the sarcoplasm from 10^{-7} to 10^{-6} mol/L, which leads to a rearrangement of troponin and tropomyosin to face myosin. The SR is a complex membranous network covering each myofibril, and its main function is to maintain a balance between Ca^{2+} and Mg^{2+} ions, controlling enzyme activity as well as controlling the levels of adenosine triphosphate (ATP), creatine phosphate (CP), and glycogen within the muscle. Excess ATP is stored in the muscle in form of CP and ADP, and muscle contraction, as well as relaxation, is an energy-consuming process.

In cases of urgent need (heavy muscular work) energy in form of ATP is provided in following the reaction: $CP + ADP \rightleftharpoons C$ (creatine) $+ ATP$. The release of Ca^{2+} ions triggers the activation of the enzyme myosin ATPase and causes

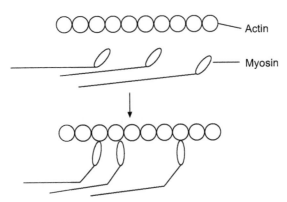

FIGURE 1.15 Binding of myosin to the actin by a change in configuration of the myosin head.

to hydrolyze ATP into ADP, phosphate, and free energy following the reaction: ATP⇒ADP+P_i+energy. The energy obtained leads to a change in configuration of the myosin head, making it bind with actin and a contracted muscle is obtained. As a result, actin and myosin are not present anymore as separate fibers but as the actomyosin complex. Within such a complex, actin and myosin are tightly bound together via cross-links (see Fig. 1.15).

As mentioned, actin undergoes a slight change in configuration so the myosin head can actually bind to the actin on a nervous impulse by Ca^{2+} ions binding to actin. Actin slides toward, or into, myosin, and therefore the degree of overlapping between actin and myosin is enhanced compared with the degree of overlap if the muscle is relaxed. This "sliding theory" is based on the fact that the width of the I-band and H-zone decreases during contraction as the thin filaments are drawn into the space between the thick filaments in the center of each sarcomere. Even by doing so, the width, or thickness, of the actin and myosin filament itself does not change during this sliding process. It is just the case that the thin filament (actin) slides deeper into the H-zone (see Fig. 1.16), and a greater degree of overlapping takes place. Due to this greater overlapping, the length of a sarcomere is shortened and the shortening of thousands of such individual sarcomeres causes the muscle to shorten overall. The length of an individual sarcomere can be reduced by up to 50% during contraction.

Relaxation of a muscle after contraction occurs via the removal of Ca^{2+} ions, and the SR absorbs the excess Ca^{2+} ions again from the sarcoplasm. The enzyme actin-myosin ATPase is activated to obtain ATP required for the separation of actin and myosin to turn the actomyosin complex into a relaxed muscle again, where both filaments are present in its separate state. The energy required for both muscle contraction and relaxation comes from the hydrolysis of ATP, and during such muscle fiber movement, chemical energy coming from ATP is converted into mechanical energy.

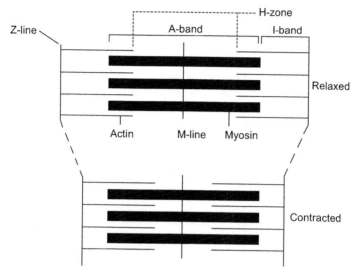

FIGURE 1.16 Relaxed and contracted sarcomere.

1.8 ENZYMES IN MEAT

Enzymes are proteins and usually have an "-ase" ending. They are generally large globular proteins, and the main characteristic of enzymes is that they can trigger, or speed up, chemical reactions within meat without being consumed during the process of the action. Enzymes act as biocatalysts by increasing the rate, or speed, of a reaction by a factor of 10^3–10^{10}. Each enzyme acts very specifically, such as splitting certain molecules only or speeding up certain chemical reactions. Also, each enzyme is characterized by specificity for a very narrow range of chemically similar substrates or reactants. The specific arrangement of an enzyme's amino acid side chain in the active site determines which type of molecule is able to react with a certain enzyme. Each enzyme has a specific three-dimensional shape with a specific surface configuration, which makes their action, or the place of action, very specific once more.

Enzyme activity is greatly influenced by pH value and temperature. Most enzymes present in meat are "working at their best" between 25 and 50°C, and activity is greatly reduced at lower pH values. The most common enzymes in meat are:

1. Lipases. Lipases are lipid-hydrolyzing enzymes, placed in the lysosomes of a cell, and is secreted by microorganisms. The enzyme breaks down fat into glycerol and free fatty acids (FFAs) in a process called lipolysis. Lipolysed fat smells and tastes rancid, and these particular smell and taste originate from low-molecular fatty acids such as butyric and caproic acid. However, a

certain degree of lipolysis is desired in long-dried products such as salami to get the product's typical slightly rancid taste. A high number of FFAs speed up rancidity in meat products. Lipase, to a great extent, is deactivated by 72°C, but 80–82°C is necessary to totally deactivate this enzyme.
2. Proteases. Several different types of proteases are present in meat, and those enzymes break down proteins into peptides as well as free amino acids in a process known as proteolysis (the ending "-lysis" refers most commonly to breakdown of substances). Proteases act in many different ways within meat and meat products, and their function as a "tenderizer" is one of the most important ones. Hence, proteolysis taking place in products such as salami has a major contribution to the typical mellow and "cheesy" flavor of such products, which is mainly based on sulfur-containing metabolic byproducts of enzyme activity.
3. Collagenases. Collagenases is able to soften collagen and specifically break down bonds within the triple helices, which form collagen; by doing so a higher number of individual molecules of tropocollagen are obtained (see Section 1.4). Hence, bonds present within a molecule of tropocollagen are loosened up as well. The bacteria *Pseudomonas aeruginosa* produces the enzyme collagenase (which is a protease), and it hydrolyzes the collagen in the connective tissue.
4. Catalase splits hydrogen peroxide (H_2O_2) into water and oxygen at an enormous speed. Around 400,000 molecules of hydrogen peroxide can be spilt within 1 second. One unit of catalase is defined as the amount of extract needed to decompose 1 μmol of H_2O_2 per minute.
Hydrogen peroxide is a very "aggressive" and slightly viscous substance, which leads to discoloration in meat products, on the one hand, and speeds up rancidity in raw fermented salami, on the other hand. Based on that, starter cultures commonly applied in the production of raw fermented salami contain catalase-positive microorganisms. The enzyme catalase is deactivated by exposing the enzyme to 74°C for 60 seconds.
5. Phosphatase can split phosphoric acid esters into phosphoric acid and the corresponding alcohol, but this enzyme plays no significant part in the production of salami.
The degradation of phosphates within meat itself and added phosphates to meat products is partly a result of the activity of the enzyme phosphatase. This enzyme is destroyed by exposing it to 72°C for 25 seconds.
6. Glycosidase is able to split glycosidic bonds between carbohydrates, which is important in case disaccharide are applied in the production of salami as starter cultures can only fermented monosaccharide in lactic acid.

1.9 FAT

Fats, or lipids, are the most concentrated source of food energy and are necessary to health. Carcass fat contains around 80–85% triacylglycerol fat, 5–10% moisture, and around 10% connective tissue. For the context of this book, fat is seen as fatty tissue given that fatty tissue and fat are by definition not the same, as "fat" refers to the fat material only without water or connective tissue. Fat is a nonpolar molecule, and unlike water, which is polar, fat does not exhibit a negative and positive end (pole). Fat, or lipids, are therefore insoluble in water due to the presence of insoluble carbon–hydrogen components within the molecule. Food fats are carriers of fat-soluble vitamins and some essential unsaturated fatty acids.

Fat is generally colorless but exhibits occasionally a touch of yellow and is by nature extremely hydrophobic (lipophilic). Fat from cattle, fed fresh grass containing carotene, frequently shows a yellow touch, and fat itself carries some flavor but is an excellent solvent for countless other flavor and aroma components. Depending on the type of fatty acids present in meat, the flavor can vary dramatically. Pork fat produces on the impact of heat saturated as well as unsaturated aldehydes, which are typical for the pork flavor. Such components are hardly present in beef fat, as beef fat contains predominantly saturated fatty acids. Some branched fatty acids are found within fat originating from sheep, which are responsible for the pronounced sheep, or lamb, flavor. The difference in flavor and taste of different types of fat is not solely based on pure fat but more due to the heat treatment of fatty tissue overall. Fatty tissue, besides fat, also contains connective tissue as well as other amino acids, which contribute to a large extend to the various flavors originating from different types of fat. Fatty acids, such as oleic acid, show a positive impact toward the flavor of fat, while stearic and linolenic acid demonstrate a negative impact on the flavor of fat overall.

Fats are divided into three major groups:

1. Intramuscular fat (fat between the muscle fiber bundles). It is also known as the marbling fat and as such plays a major role toward juiciness, flavor, and tenderness of meat. The world famous Kobe beef in Japan has extremely high levels of such marbling fat, which, beside other factors such as a very special diet and treatment of the animal overall, contributes to the very tender, juicy, and tasteful meat.
2. Intermuscular fat (between individual muscles)
3. Subcutaneous or depot fat (under the skin)

The building blocks for fat (or simple lipids) are triglycerides, and fat is made of molecules of carbon, hydrogen, and oxygen. Other substances, known as complex lipids, also contain phosphorous, nitrogen, and sulfur, besides carbon, hydrogen, and oxygen. Triglycerides in fat are esters of the trihydric alcohol glycerol, and three fatty acids (three of the same type or three different ones) are bound to glycerol. An ester is a compound formed from the reaction between an alcohol and acid by the removal of water. The reaction is the following: Alcohol (glycerol) + acid (fatty acids) $\Rightarrow$ ester + water (see Fig. 1.17).

Glycerol (or 1,2,3-propane triol) is an alcohol showing three OH groups within its molecule. When triglycerides are solid at room temperature, they are called "fats." On the other hand, in case triglycerides are liquid at room temperature, they are called "oils."

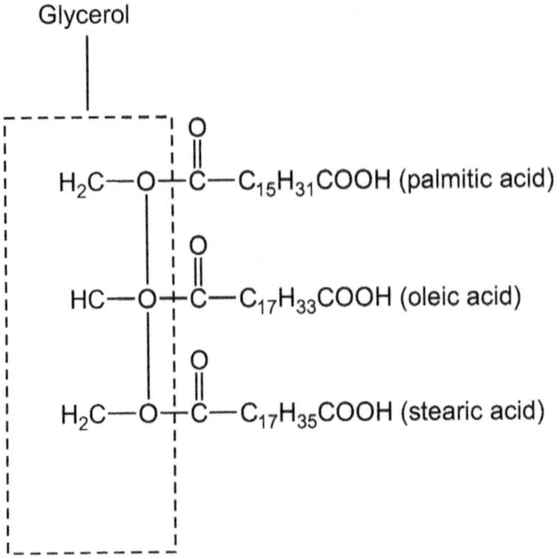

```
                    Glycerol
                       |
      ┌ ─ ─ ─ ─ ─ ─ ─ ─┼ ─ ─ ─ ─ ┐
      |                   O      |
      |                   ‖      |
      |    H₂C—O—┬—C—C₁₅H₃₁COOH (palmitic acid)
      |          |                |
      |          |        O       |
      |          |        ‖       |
      |    HC—O—┼—C—C₁₇H₃₃COOH (oleic acid)
      |          |                |
      |          |        O       |
      |          |        ‖       |
      |    H₂C—O—┴—C—C₁₇H₃₅COOH (stearic acid)
      |                           |
      └ ─ ─ ─ ─ ─ ─ ─ ─ ─ ─ ─ ─ ─ ┘
```

FIGURE 1.17 Molecule of triglyceride consisting of the alcohol glycerol and three fatty acids.

Lipids include mono-, di-, and triglycerides; sterols; terpenes; phospholipids; fatty alcohols; and fatty acids. Phospholipids, such as lecithin, exhibits two fatty acids and a phosphoric component bound to glycerol. Cholesterol is the most well-known representative from the sterols group. Monoglycerides have one fatty acid bound to glycerol, while diglycerides demonstrate two fatty acids bound to glycerol. Triglycerides exhibit three fatty acids bound to the alcohol glycerol.

Fatty acids consist of long chains of hydrocarbon with a carboxyl group (—COOH) at the end. The carboxyl group (—COOH) is carbon 1 for the purposes of naming the fatty acid. This group is shown in a chemical formula mostly at the right-hand end, while the methyl (—CH_3) end is generally shown on the left-hand end of a fatty acid. Counting of the total carbon atoms starts from the COOH carboxyl end. For example, the fatty acid C18:2 contains 18 carbon atoms, and two double bonds are present within the fatty acid. Numbers such as 9 and 12 are also shown in conjunction with the name of the fatty acid, and this indicates that two double bonds are present within this fatty acids, and those double bonds are located at carbon atom numbers 9 and 12, counted from the carboxyl end.

Stearic acid is a C18:0 saturated fatty acid exhibiting the carboxyl end as well as the methyl end. The term C18:0 means that the fatty acid contains 18 atoms of carbon, and no double bond is present. Hence, the α-carbon is the first, or the closest, carbon to the carboxyl group (COOH) and is theoretically the second carbon in the chain from the COOH group. As a result, stearic acid can also be expressed as $C_{17}H_{35}COOH$ based on the fact that all carbon atoms are counted except the carbon belonging to the carboxyl group (COOH). Counting

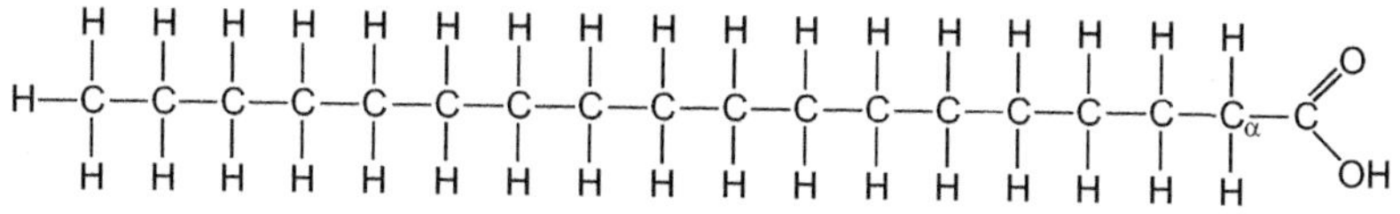

FIGURE 1.18 Stearic acid.

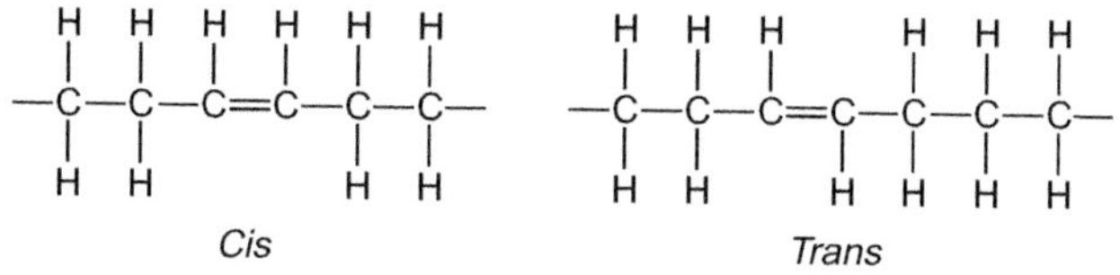

FIGURE 1.19 The *cis-* or *trans*-configuration in fatty acids.

therefore starts at the α-carbon. The most common fatty acids also have a scientific name. Stearic acid (C18:0) is also known as octadecanoic acid, oleic acid (C18:1) is known as 9-octadecenoic acid, and linoleic acid (C18:2) is known as 9,12-octadecadienoic acid (see Fig. 1.18).

Double bonds between carbon atoms stabilize the structure of fatty acids by preventing the carbons from rotating around the bond axis. As a result, configurational isomers of the same fatty acid are obtained, and the arrangement of atoms within such fatty acids can only be changed by breaking the bonds.

This fact gives rise to either *cis-* or *trans*-configuration of fatty acids, and the Latin prefixes *cis* and *trans* demonstrate the location of the hydrogen atoms in respect of the double bond. The term *trans* indicates the other, or opposite, side, while *cis* means on the same side. Fatty acids generally exhibit *cis*-configuration, while the *trans*-configuration is more often present in nature overall (see Fig. 1.19).

Fatty acids, demonstrating the same number of carbon atoms as well as number of double bonds at the same carbon number, become different fatty acids depending on whether *cis-* or *trans*-configuration is present.

The different types of fats are:

- Saturated fats
- *trans*-Fats (behave similar to saturated fats)
- Monounsaturated fats
- Polyunsaturated fats

The saturation of fat refers to the chemical structure of its fatty acids. Saturated fatty acids are of linear structure (nonbranched) and generally exhibit even numbers of carbon atoms within their molecule such as 16 or 18 carbons. Single-bond linkages are present in saturated fatty acids between carbon atoms, and no double bond is given. Such single-bond linkages are chemically not very active, and saturated fatty acids are commonly solid at room temperature. Animal fats are predominantly saturated fats or contain a high amount of saturated fatty acids. Important representatives of saturated fatty acids present in animal fat are stearic acid (C18:0) as well as palmitic acid (C16:0). The "0" indicates that no double bond is present within the fatty acid, which is made of 18 atoms of carbon. Beef fat contains a high level of saturated, long-chained fatty acids. The degree of saturation in fat decreases in the sequence beef>pork>poultry>fish (least saturated). Stearic acid is unique in a way that it does not raise blood cholesterol and unfortunately is very often associated with other saturated fatty acids, which do raise blood cholesterol. Major sources of stearic acid are chocolate, lard, tallow, and commercial fats and butter. Palm and coconut oils are also rich in saturated fatty acids.

Unsaturated fatty acids contain one or more double bond(s) between carbon linkages, and the double bonds in unsaturated fatty acids regularly show *cis*-configuration. Monounsaturated fat is a type of fat in which the fatty acid contains one double bond in its chemical structure. Such fatty acids are found in olive, canola, and peanut oil and avocados. Oleic acid is a C18:1 monounsaturated fatty acid, which exhibits one double bond after the ninth carbon from its carboxyl end (—COOH) and shows 18 carbon atoms within it molecule (see Fig. 1.20).

Monounsaturated fats can lower the total cholesterol by replacing saturated fats and do not lower the level of the "healthy" high-density lipoprotein cholesterol. They are less prone to oxidation compared with polyunsaturated fats. Fats that contain monounsaturated fatty acids are normally liquid at room temperature, but many thicken up when placed under refrigeration. Monounsaturated fats are beneficial to health and may be better than polyunsaturated fats in preventing heart disease. The diet in countries such as Italy and Greece is high in monounsaturated fats coming from olive oil and is one explanation for the low rate of heart disease in those countries. Olive oil contains around 75% oleic acid (C18:1).

Polyunsaturated fatty acids exhibit two or more double bonds within their molecule and the two main types are:

1. Omega-3 fatty acids such as alpha-linolenic acid (ALA), which is the starter fatty acid for the omega-3 series. This fatty acid is 18 carbons long and shows in total three double bonds placed after the third, sixth and ninth carbon from the methyl end (—CH₃) within the molecule. Other representatives in this group are docosahexaenoic acid (DHA) and eicosapentaenoic acid (EPA).

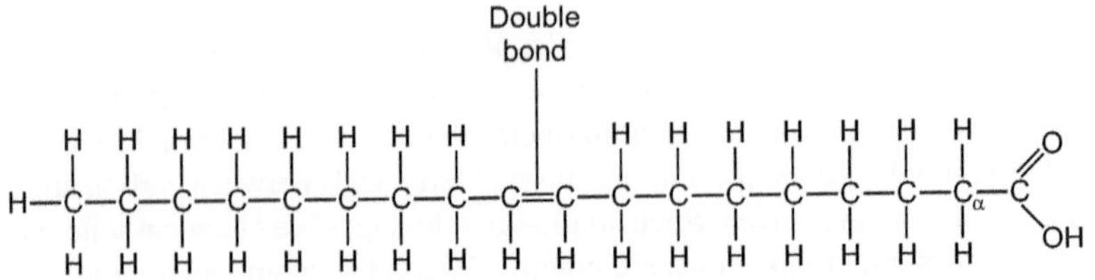

FIGURE 1.20 Oleic acid.

2. Omega-6 fatty acids such as linoleic acid, which is the main polyunsaturated fatty acids in vegetable oils, originate from canola, maize, sunflower, or peanut. Linoleic acid is the starter fatty acid for the omega-6 series, showing 18 carbon atoms as well as two double bonds placed after the carbon atom number six and nine from the methyl end within the molecule. Other members of this group are gamma-linoleic acid (GLA) and arachidonic acid (AA). Omega-9 fatty acids, such as palmitoleic acid, also exist (see Fig. 1.21).

Omega-3 and omega-6 fatty acids are essential unsaturated fatty acids and have to be provided to the human body by eating foods containing those fatty acids given that the human body cannot build them out of other fatty acids. Polyunsaturated fatty acids are named by counting from the methyl ($-CH_3$) end within the fatty acid. Letters from the Greek alphabet such as alpha, beta, gamma to omega are utilized in order to determine the location of double bonds within the polyunsaturated fatty acid. Omega is the last carbon in the chain of carbon atoms counted from the carboxyl COOH end as the letter omega is the last letter in the Greek alphabet. For example, linoleic acid is an omega-6 fatty acid and the first double bond, counted from the methyl end, is located six carbons away from the omega carbon. This omega-carbon is the same carbon as used within the methyl (CH_3) group and carbon number 18 from the carboxyl end. Polyunsaturated fats, such as corn oil, are generally liquid at room temperature as well as under refrigeration (see Fig. 1.22).

Different types of fat exhibit different melting points and as a result have a different impact on the mouth feel in meat products. Fats, containing a high

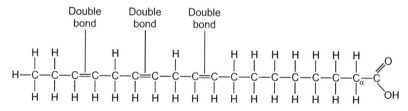

FIGURE 1.21 Alpha-linolenic acid.

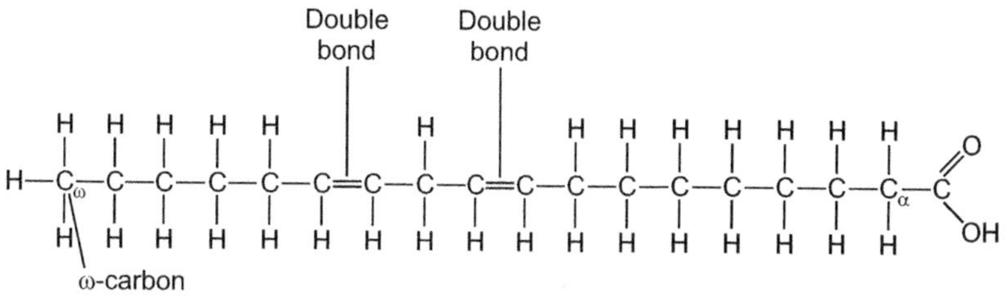

FIGURE 1.22 Linoleic acid.

number of saturated fatty acids (like kidney fat or lard) cause a greasy, smeary and sandy mouth feel while more unsaturated fats give a pleasant taste as well as smooth, nonsandy mouth feel. Generally, the "hardest" fat within a carcass, showing high levels of saturated fatty acids, are found in the center of the carcass and softer fats are placed toward the outside of a carcass.

Even in subcutaneous pork fat, such as pork back fat, the outer layer of pork back fat, directly connected with the skin, is softer compared to the inner layer. Hence, soft fat contains a higher amount of connective tissue within itself compared with hard fat and chicken fat, which is the softest fat (high amount of unsaturated fatty acids). On the other hand, beef fat, which is of hard consistency, exhibits the lowest level of connective tissue. Pork fat lies between chicken and beef fat in terms of the level of connective tissue within the fat itself. In summary, soft fat contains higher levels of connective tissue but fat molecules entrapped within connective tissue are of soft consistency (high degree of unsaturated fatty acids). Hard fat, on the other hand, contains less connective tissue but fat molecules covered by connective tissue are of hard consistency (high degree of saturated fatty acids).

The melting point of a fatty acid depends largely on the length of the fatty acid itself as well as the number of double bonds present. Saturated fatty acids generally show a higher melting point compared with unsaturated fatty acids. Double bonds (and therefore less hydrogen within a fatty acid), present in unsaturated fatty acids, lower the melting point and unsaturated fatty acids show generally a lower melting point compared to saturated fatty acids as a result. An increased number of double bonds within a fatty acid lowers the melting point once again. Increased length of a fatty acid, containing a higher number of carbon atoms, causes an increase in melting point. For example, stearic acid (18 C atoms) has a melting point around 70°C, while capric acid (10 C atoms) shows a melting point around 30°C. Overall, the melting point of beef fat is around 43–47°C, pork fat around 38–44°C, and chicken fat around 31–37°C. Hence, fat containing *cis*-shaped double bonds within the fat molecule exhibits a lower melting point compared to fat containing *trans*-double bonds.

The consistency of fat is largely depended on the saturation of the fatty acids.

A higher number of unsaturated fatty acids leads to "softer" fat. Pork fat contains a relatively high amount of unsaturated fatty acids and is "soft" as a result. Beef fat, on the other hand, contains predominantly saturated fatty acids and therefore is of a "hard" consistency. The level of saturated fatty acids within fats varies and is for beef around 55–60%, for pork around 42–44%, and for chicken only 30%. This explains the "hardness" of fat in the sequence beef⇒pork⇒chicken, with chicken being the softest. Lamb and mutton are similar to beef in regard to the content of saturated fatty acids. For the production of meat products, pork fat showing a low number of unsaturated fatty acids, such as fat from loin and neck, is the preferred choice over soft pork fat, coming from leg and shoulder, showing a higher number of unsaturated fatty acids.

Such "soft" fat is best used for emulsified sausages and not recommended for products such as salami, where "hard" fat is needed as it can be cleanly cut and the tendency toward rancidity is reduced as well.

1.10 RANCIDITY OF FAT

Unsaturated fatty acids, showing double bonds within the molecule, are susceptible to peroxidation. Such peroxidation could be called "oxidative deterioration" of lipids and can occur via a radical reaction in two ways. One way, and by far the most important one, is autoxidation. The other way, less important, is an enzymatic oxidation. Oxidation is a process where an oxygen ion replaces a hydrogen ion within a fatty acid molecule and higher numbers of double bonds within the fatty acid increase the possibility of autoxidation. Pork and chicken fat demonstrate a higher degree of unsaturated fatty acids compared with beef fat and are therefore more prone for rancidity.

The availability, or presence, of oxygen, increased temperatures, impact of light, and the presence of prooxidants speed up autoxidation (oxidative rancidity) over a period of time. Autoxidation does not take place from one day to another, but the potential for the chain reaction to start increases over time and, once triggered, takes place at a fast rate. The presence of prooxidants such as iron and copper ions accelerates the onset of autoxidation, as does the presence of oxygen and exposure to light, especially direct sunlight or light from fluorescent tubes. Iron and copper ions can be deactivated by the addition of chelating agents such as citric acid. The oxidation of fat occurs at a faster rate at a reduced water content given the fact that water acts as a "barrier" against the reaction of fatty acids with oxygen. The smaller the quantity of water within food, the more "effective" the oxygen is toward oxidation.

Autoxidation is an oxygen-induced radical chain reaction and can be divided into the three phases of initiation, propagation, and termination. During the initiation phase, free radicals are obtained. Those free radicals react with other materials during the propagation phase, and nonradical products are obtained during the phase of termination. It was thought in the past that oxidation occurs directly on the double linkages between two carbon atoms within fatty acids, but it is known today that a double linkage between two carbon atoms favors the oxidation of the neighboring carbon linkage. The placement of oxygen on a carbon linkage within an unsaturated fatty acid is known as peroxidation. The removal of hydrogen from the fatty acid, through the impact of oxygen, causes the formation of a fatty acid radical as well as a hydrogen radical. Through such a process, an oxygen ion replaces a hydrogen ion within a fatty acid (initiation phase).

A radical is a molecule, or part of a molecule, with a nonsaturated pair of electrons and therefore a highly reactive substance. Radicals are rich in energy and are generally obtained from a nonradical losing one electron and is symbolized in a chemical formula with a dot such as [R˙] or [R*].

As said, radicals are extremely reactive due to the lack of an electron and, being in such an "unstable" state, the radical tries to stabilize itself by obtaining an electron from another molecule. The fatty acid radical reacts with other fatty acids and cause the formation of new radicals and the removal of an electron from another nonradical material to stabilize the original radical causes the formation of another radical. As a result, a self-feeding chain reaction is started. On the availability of oxygen, radicals bind with oxygen and activated peroxide radicals are the result. Such peroxide radicals are stabilized via the binding of hydrogen atoms, originating from other fatty acids, and hydroperoxides are the result. All those reactions of radicals with other substances take place during the propagation phase.

A hydroperoxide is by itself a nonradical intermediate but creates within the ongoing process of autoxidation one radical and one peroxide radical.

Hydroperoxides are also neutral from a sensorial point of view but demonstrate an extremely high potential for oxidation. They ultimately fall apart in the termination phase into countless relatively unreactive components including aldehydes, hydrocarbons, and ketones but also, as said before, new radicals (highly reactive). Some of the aldehydes obtained from hydroperoxide is malone dialdehyde and its isomeric combinations, hydroxyacrolein and ephidrinaldehyde. Aldehydes and ketones, originating from hydroperoxides, are primarily responsible for the rancid smell of fat. Both of those substances are well suited for testing fat regarding its state of rancidity. The self-feeding chain reaction as well as the fact that hydroperoxides create new radicals explains that oxidation cannot be totally stopped once it started—it only can be delayed.

Up to 100 peroxides, or hydroperoxides, can be formed out of one radical, which explains why once autoxidation has started, it continues in an exponential way within the propagation phase.

Hydrolytic rancidity based on enzymatic oxidation is another radical-based reaction but of much less significance than oxidative rancidity. During this process, lipases create FFAs, which are prone afterward toward oxidation. Such hydrolytic rancidity can also get started by the reaction between water and lipids. The application of antioxidants, or storage of fatty meat (or fat) at $-18°C$, does not stop the process of fat going rancid. Some lipase exhibit activity even at temperatures as low as $-28°C$.

Photo-oxidation is another form of oxidation, and oxygen changes its configuration from triplet oxygen $\left(^3O_2\right)$ to singlet oxygen $\left(3 \times {}^1O_2\right)$ during the process through the impact of short-wave UV light. Singlet oxygen is significantly more reactive compared with triplet oxygen and singlet oxygen $\left({}^1O_2\right)$ reacts with unsaturated fatty acids to form hydroperoxide, which is in turn broken down to free radicals, and this chain reaction starts again. Within autoxidation, the oxygen involved never changes its configuration, which is a distinctive difference to photo-oxidation, where oxygen involved changes to singlet oxygen.

The presence of light also demonstrates a large impact toward obtaining and speeding up the process of rancidity. It is well known that fatty meat (or fat), stored in darkness at freezing temperatures, develops rancidity at a much slower rate than if the same materials would be stored at the same temperatures under the impact of light. UV light supports the formation of radicals and therefore speeds up rancidity. Different types of fat develop rancidity at different speeds. Fatty pork meat, or pork fat, should not be stored longer than 3–4 months at −18°C, while 6 months is the maximum for fatty beef, or fatty lamb meat, stored at −18°C. Lean meat, due to the low content of fat, can be stored up to 1 year at −18°C.

Rancidity is measured the following ways:

1. Number of peroxides
   Peroxide values (PV number), or oxidative rancidity, measures the number of peroxides present in fat. Therefore, fat has to be isolated and other components, such as protein, removed to the outmost extent as those nonfat components can interfere during the analysis, which is a titration process. Measuring the number of peroxides is also commonly done on pure fats and oils and shows the stage of oxidation, or how far oxidation has progressed, as peroxides are the first components formed during the oxidation of fat. The PV value is helpful in order to determine the quality of saturated fat but is not that helpful in unsaturated fats. Peroxides formed in unsaturated fats such as pork and chicken fat contain a large amount of unsaturated fatty acids, are unstable, and quickly react further into secondary oxidation substances. Therefore, getting an accurate reading on the number of peroxides is difficult. Peroxide itself does not show sensory characteristics that are related to rancidity but are intermediate substances that react further to ketones and aldehydes, the main odorous substances in rancid fat. As a result, the number of peroxides found in unsaturated fat can be low even though the fat is already in a high state of oxidation. Even severely rancid fats can exhibit a low number of peroxides due to those secondary reactions.
   The number of peroxides and hydroperoxide, generally formed in the early stage of oxidation, is expressed in milliequivalents of peroxide per kg of fat (mEq/kg). The formation of peroxide takes place at a slow rate during initial stages of oxidation depending on storage temperature and the presence of antioxidants, but once a "critical mass" is obtained, the increase in peroxides takes place exponentially. PV numbers from 0 to 6 are generally seen as fat not being rancid, while numbers from 7 to 10 are seen as fat being slightly rancid. Numbers greater than 10 clearly indicate rancidity but keep in mind that the PV number does not always directly relate to the state of rancidity (as explained earlier).

2. Determination of the thiobarbituric acid value (TBA value)
   This method is widely applied in meat products, and the TBA value equals milligrams of malonaldehyde/kg of sample. The amount of malonaldehyde is determined in a photometric way, and rancidity starts at 0.5–0.8 mg of

malonealdehyde/kg of sample. Within this test, saturated aldehydes, obtained during the termination phase of fat oxidation, react with 2-TBA. TBA value analysis is done on the food overall and not on fat, only in contrast to peroxide number or FFA tests (see later). TBA numbers generally correlate with the state of rancidity and increased numbers indicate an advanced state of rancidity.

3. Analyzing the FFA content is based on measuring hydrolytic rancidity in fat or oil. This type of analysis is done on fat only, and hydrolytic rancidity originates from the hydrolysis of triglycerides in the presence of moisture. Enzymes such as lipase generally speed up this process, and the hydrolysis results in FFAs. FFA numbers generally increase during storage of fat, or fatty meats, but oxidative rancidity has a greater impact toward rancidity as hydrolytic rancidity. Hydrolytic rancidity is expressed generally within fat from meat or meat products in percent oleic acid and other fats, showing a high degree of shorter-chain fatty acids (coconut oil) as percent lauric acid. FFA numbers in meat and meat products above 1.2 indicate rancidity.

Other, not commonly applied methods, to determine rancidity, is the determination of the hexanal value, the total oxidation volume, and the oxidative stability index test. Hexanal is an aldehyde, produced during the termination phase of fat oxidation, and is measured via a distillation process or by carrying out a gas chromatographic analysis of the headspace over a sample. The amount of hexanal determined correlates with sensorial spoilage and a concentration beyond 6 µg/kg indicates rancidity.

In determining the total oxidation volume, which is predominantly applied by pure fats and oils, a 5-g sample is exposed to oxygen until no more oxidation takes place within the sample. The total amount of oxygen utilized during the process is expressed in mEq/oxygen per 5 g of fat or oil and is determined in a volumetric way. Another predictive test toward fat oxidation would be the oxidative stability index test.

Chapter 2

Biochemistry of Meat

2.1 BIOCHEMICAL PROCESSES IN MEAT PRESLAUGHTER

In preslaughter muscle tissue, as long as the animal is alive and therefore breathing, muscle contraction and relaxation can take place in an aerobic way by the presence of oxygen. During muscle movement, the filaments actin and myosin slide into each other and chemically link and unlink causing the muscles to contract as well as relax. Energy in the form of adenosine triphosphate (ATP) is required to bind myosin into actin as well as to separate myosin from actin afterward.

The formation of ATP, the ultimate source of energy, is a highly complicated process and ATP is normally present in muscle tissue at a level of around 5–6 μmol/g of tissue. A molecule of ATP consists of a D-ribose (sugar), adenine, and three phosphate groups. Similar compounds are adenosine diphosphate (ADP) and adenosine monophosphate (AMP), which show two phosphate groups or only one phosphate group within the molecule, respectively (Fig. 2.1).

ATP hydrolyzes easily to ADP, a single phosphate unit and energy. The chemical energy obtained is used for all energy-consuming processes within the living organism as well as for the movement of muscle fibers. Excess glucose is stored in muscle tissue and in the liver in form of glycogen as well as creatine phosphate (CP) and is converted into energy if required. Hence, if all excess energy were to be stored in the form of ATP directly, the muscle would be in a state of permanent contraction. CP is required in times where insufficient ATP is formed or during times where ATP has to be provided instantly. ATP can be quickly synthesized in such situations and CP binds quickly with ADP to form ATP [CP+ADP⇒C (creatine)+ATP].

The formation of ATP while the animal is still alive and breathing can take place via the use of glucose, proteins, or fat. All of those materials can be transformed into acetyl coenzyme A (acetyl-CoA), which is the substance that enters the citric acid cycle. Carbohydrates, however, are primarily used for rebuilding ATP, and glycogen is readily broken down to glucose when required for energy. Fat or proteins are used for the formation of ATP when no more carbohydrates are available.

The process of rebuilding of ATP can be divided into three steps:

1. Glycolysis is the oxidation of glucose to pyruvate with the formation of some ATP and energy-rich coenzymes such as NADH (NADH dehydrogenase) from nicotinamide adenine dinucleotide (NAD^+). NADH dehydrogenase is a flavoprotein (enzyme) that contains iron-sulfur centers.

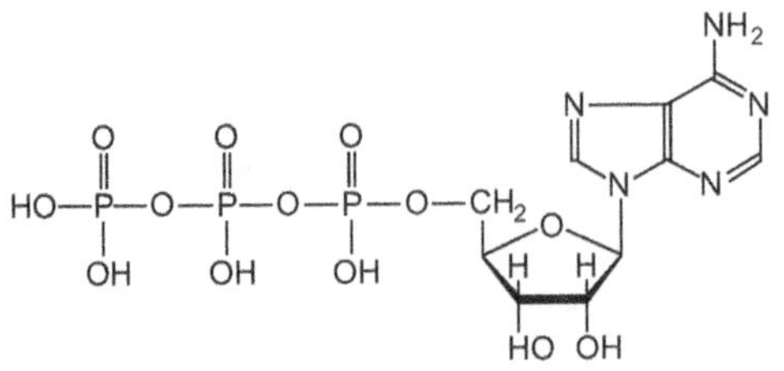

FIGURE 2.1 Adenosine triphosphate.

2. Within the second step, pyruvate is transformed into acetyl-CoA, which enters the Krebs or tricarboxylic acid cycle and is ultimately oxidized into CO_2 and water. Within this process, substances such as guanosine triphosphate (GTP) as well as energy-rich reduced coenzymes like NADH and $FADH_2$, which is formed out of flavin adenine dinucleotide (FAD), are obtained.
3. Third, NADH and $FADH_2$ are oxidized within the oxidative phosphorylation. The passing-on of electrons from one carrier to another results in the formation of free energy, which is used to synthesize ATP.

2.1.1 Glycolysis

Glycogen, the muscular sugar and a polysaccharide, has a molecular mass of 6–12 MDa, and muscle tissue contains around 0.7–1.1% glycogen. Muscle tissue also contains around 10–25 g of free glucose per 100 g of tissue. Glycogen is only found in animals and can be seen as the equivalent storage carbohydrate to starch in plants. Glycogen is a polymer of D-glucose and is identical to amylopectin found in starch, but the branches in glycogen tend to be shorter (about 10–12 glucose units) and more frequent than in amylopectin. The glucose chains are organized globularly, and glycogen can be converted back to glucose quickly. The two forms of glycogen, commonly referred to as glycogen for reasons of simplicity, are pro- and macro-glycogen. Pro-glycogen accounts for around 75% of the total glycogen, while macro-glycogen makes up the remainder.

Glycogen is converted easily into glucose, which is then broken down by glycolysis, also known as the Embden–Meyerhof pathway. The process of glycolysis is the oxidation of glucose to pyruvate, which takes place in the sarcoplasm (cytoplasm) of a cell and is heavily regulated by enzymes. It is an anaerobic process and, therefore, no oxygen is required. During glycolysis, glucose is turned into pyruvate by first being turned into glucose 6-phosphate. Glucose-6-phosphate is then transformed into fructose-6-phosphate; 10 steps are needed in total to obtain pyruvate, which is the end point of glycolysis.

One molecule of glucose entering glycolysis turns into two molecules of pyruvate as glucose is a six-carbon molecule and pyruvate is a three-carbon molecule. Pyruvate is finally converted into acetyl-CoA by the loss of one molecule of CO_2 to become a two-carbon molecule in a process known as decarboxylation. This transformation is supported by the enzyme pyruvate dehydrogenase, and acetyl-CoA is occasionally referred to as "activated acetic acid." Splitting of glucose during glycolysis is an exergonic process, meaning that energy is released during the process. Within glycolysis, two molecules of ATP are used but four molecules of ATP are gained. Therefore, the net result of glycolysis is +2 ATP per molecule of glucose.

When there is an oxygen deficiency during periods of instant need of high levels of ATP (such as heavy muscular work or in a very stressful situation such as the time before slaughter), ATP is obtained anaerobically and lactic acid is formed as a byproduct in the living organism. This organic acid is then transported in the bloodstream back to the liver, where lactic acid is converted first into glucose-6-phosphate and then into glycogen or glucose.

Proteins are generally not used to obtain ATP, but in times of great need, such as when further carbohydrates are not available, proteins are also converted into acetyl-CoA. Within such very rare situations, proteins are broken down in the first place into amino acids, which then are converted in a second step into acetyl-CoA. This process is called deamination and the amino group of an amino acid is removed. Subsequently, the amino group is converted into an ammonium ion (NH_4^+), which is removed from the cell. The remaining part of the amino acid can then enter the Krebs cycle. Proteins are, as said, the last resource for obtaining energy, given that proteins are too valuable to be transformed to energy. The body uses carbohydrates (and fat) first, whereas proteins are the very last resort.

In cases where fat is broken down for the formation of ATP, glycerol is turned into aldehyde phosphate together with fatty acids, which are then oxidized several times, also ending up as acetyl-CoA. The formation of pyruvate is the end point of anaerobic glycolysis in the cytoplasm using glucose before pyruvate is turned into acetyl-CoA. In effect, all materials such as amino acids (from proteins), fatty acids and glycerol (from fat), and pyruvate (from glycogen) end up in the form of acetyl-CoA, but by far the most important material is glucose. Proteins and fat are transformed directly to acetyl-CoA, whereas glucose is, as said earlier, transformed first into pyruvate before then being turned into acetyl-CoA.

2.1.2 Krebs Cycle

This biochemical cycle, also known as the tricarboxylic acid cycle (TCA) or citric acid cycle, is named after the Austrian scientist Hans Krebs, who researched this highly complex process. The citric acid cycle is a sequence of reactions in which two-carbon molecules of acetyl-CoA, originating from pyruvate (or proteins and fat), are entirely oxidized to CO_2 and water. This biochemical process occurs in the mitochondria of a cell and oxygen is required for this aerobic process. Free hydrogen atoms are obtained during the citric acid cycle, which bind to coenzymes such as NAD^+ and are then reduced to NADH as a result. Hence, other coenzymes such as FAD also take up free hydrogen and are reduced to $FADH_2$. Those reduced coenzymes contain more energy than in their nonreduced state, and this energy is used in the last step of oxidative phosphorylation to synthesize ATP out of ADP.

The major steps within the citric acid cycle are:

- Catalysation of acetyl-CoA to citrate by the help of the enzyme citrate synthase
- Catalysation by aconitase, another enzyme, of citrate into isocitrate
- Decarboxylation of isocitrate into α-ketoglutarate by isocitrate dehydrogenase as well as reducing NAD^+ to NADH
- Production of succinyl-CoA by α-ketoglutarate dehydrogenase and reduction of NAD^+ to NADH within this step
- Transformation of succinyl-CoA into succinate by the enzyme succinyl CoA-synthetase, obtaining GTP as well
- Oxidation of succinate to fumarate by succinate dehydrogenase and reduction of FAD to $FADH_2$
- Hydration of fumarate to malate by fumarase
- The last step within the citric acid cycle: transformation of malate into oxaloacetate by the help of the enzyme malate dehydrogenase by another reduction of NAD^+ to NADH (Fig. 2.2)

For every two molecules of acetyl-CoA entering the Krebs cycle, four molecules of CO_2 are liberated via decarboxylation. In addition, two molecules of $FADH_2$, six molecules of NADH, as well as two energy-rich molecules of guanosine triphosphate (GTP) are obtained. The formation of the coenzymes NADH and $FADH_2$, out of NAD^+ and FAD, is the most important process of the Krebs cycle as these reduced coenzymes contain a lot of energy, which is used in a series of reduction as well as oxidation processes within the subsequent

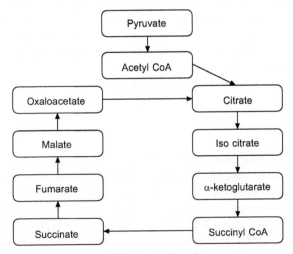

FIGURE 2.2 Krebs cycle.

step of oxidative phosphorylation for the rebuilding of ATP. One molecule of NADH results in the formation of three molecules of ATP, while one molecule of $FADH_2$ generally results in the formation of two molecules of ATP.

2.1.3 Oxidative Phosphorylation

The third step within the entire process of rebuilding ATP is oxidative phosphorylation, where reduced coenzymes such as NADH and $FADH_2$ are oxidized by the help of oxygen: this is why animals and humans have to respire. Oxidative phosphorylation also takes place in the mitochondria of a cell. Hydrogen is split into protons and electrons, and electrons are passed on to oxygen and other inorganic compounds. In a series of reactions, electrons are passed on from one carrier to another within the process of oxidative phosphorylation in what is known as the electron transfer chain (ETC). Water is split off during this process, and the reoxidation of NADH, as well as $FADH_2$, or the transfer of electrons from one electron carrier to the next one, releases energy, which is ultimately used for the formation of ATP out of ADP and phosphate (P). The process of ATP synthesis using "free energy" obtained through the passing-on of electrons to several carriers (ETC) is known as chemiosmosis. The actual point of the synthesis of ATP takes place when electrons pass the inner mitochondrial membrane. Energy is released within this process, resulting in the synthesis of ATP. Oxidative phosphorylation could be summarized in the following way:

Reduced coenzymes $NADH/FADH_2$ + oxygen $\Rightarrow$ ETC $\Rightarrow$ oxidized coenzymes NAD^+, FAD + water + free energy $\Rightarrow$ ADP $\Rightarrow$ ATP.

Thirty-two molecules of ATP are obtained during oxidative phosphorylation per molecule of glucose. Together with the two molecules of ATP resulting from glycolysis, as well as the two molecules of GTP from the Krebs cycle (which can be readily transformed into ATP and can be counted as ATP), 36 molecules of ATP are obtained in total per molecule of glucose.

Some enzymes can only function in conjunction with coenzymes. Coenzymes are not protein based and can be nucleotides, ions, or vitamins, and when these nucleotides, ions, or vitamins are bound loosely to an enzyme, a coenzyme is obtained. One of the most important tasks of a coenzyme is the carrying-over, or passing-on, of hydrogen, electrons and energy within biochemical processes. Substances, such as nucleotides, act as carrier-materials for hydrogen and electrons. Nucleotides such as NAD^+ and FAD, both coenzymes, are reduced by the uptake of hydrogen resulting in NADH as well as $FADH_2$.

As explained earlier, in preslaughter muscle tissue, as long as the animal is alive and therefore breathing, the filaments actin and myosin slide into each other, and muscle contraction, as well as relaxation, takes place promoted by aerobically formed ATP. Availability and utilization of ATP break down the actomyosin complex, obtained during contraction, into the separate fibers of actin and myosin once again. The pH value of lean muscle tissue at this stage,

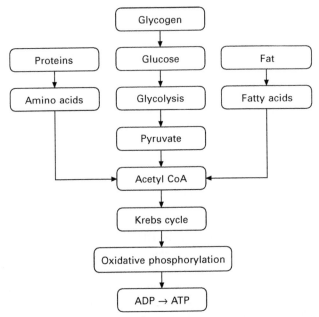

FIGURE 2.3 Process of obtaining adenosine triphosphate under aerobic circumstances.

in the living animal, is between 6.8 and 7.2. Pork fat exhibits a pH of 6.3–6.6 at point of slaughter, and beef lard, a pH value of around 6.8 (Fig. 2.3).

2.2 BIOCHEMICAL PROCESSES IN MEAT POSTSLAUGHTER (RIGOR MORTIS)

Postslaughter, chemical changes in muscles tissue due to the nonpresence of oxygen result in a situation where actin and myosin are not present as separate fibers any longer but bound together in the actomyosin complex. Muscle tissue ultimately remains contracted and processes leading to that state are known as rigor mortis. Rigor mortis, coming from Latin meaning "stiffness of death," eventually subsides to relaxation of muscle tissue once insufficient energy is rebuild and the fact that energy present in muscle tissue at point of slaughter is primarily decomposed to lactic acid. The biochemical processes in meat postslaughter leading to rigor mortis and subsequent stages are explained next.

2.2.1 Anaerobic Glycolysis and the Formation of Lactic Acid

Once the animal is slaughtered, because respiration ceases, no additional oxygen enters the body. Muscular sugar glycogen, present at point of slaughter in muscle tissue, is converted anaerobically after slaughter into pyruvate, just as it is during glycolysis, while the animal is still alive, as this step occurs anaerobically regardless of whether the animal is alive or dead and the nonpresence (or presence) of oxygen does not play a role at this point. Further, any remaining CP in muscle tissue at point of slaughter is also used during post mortem glycolysis to turn ADP into ATP under anaerobic conditions. Most of the post mortem glycolytic enzymes are located near the F-actin, which is part of the I-band.

The major difference between aerobic and anaerobic glycolysis is that the pyruvate obtained from glucose is not converted into acetyl-CoA in anaerobic glycolysis as the oxygen required for this step is no longer available. As a result, no acetyl-CoA enters the citric acid cycle to create reduced coenzymes such as NADH and $FADH_2$, and oxidative phosphorylation also does not take place. Instead, pyruvate is reduced predominantly to lactic acid (lactate$^-$ and H$^+$) under anaerobic circumstances, and this is catalyzed by the enzyme lactate dehydrogenase.

The level of glycogen in muscle tissue prior slaughter is around 7–11 g/kg, and lactic acid is formed at an amount of 38 molecules per molecule of glucose in post mortem glycolysis during rigor mortis. The lactic acid formed is not transported back to the liver, as occurs while the animal is still alive, and therefore the concentration of lactic acid within muscle tissue increases steadily after slaughter. As a result, the pH of muscle tissue declines during rigor mortis by around 1.5–1.7 pH units due to the accumulation of lactic acid from approximately 7.0 (in the living animal) down to around 5.4.

A tiny amount of ATP, however, is still obtained anaerobically postslaughter. Only three molecules of ATP are obtained out of one molecule glucose during post mortem glycolysis, compared with 36 molecules of ATP during aerobic glycolysis, representing just one-twelfth compared with aerobic glycolysis. The effects of the three molecules of ATP, obtained in an anaerobic state, can be observed on the carcass as there is visible fiber movement even though the animal is already dead.

Post mortem glycolysis can be inhibited, or comes to an end, in two possible ways. First, when no more glycogen or glucose is present within the meat, no further pyruvate is formed and therefore cannot be subsequently turned into lactic acid. Second, when during glycolysis the pH value has declined to around 5.3–5.4, even though some glucose is still present in the muscle tissue, the glycolytic enzymes cease to function at such low pH values. Anaerobic glycolysis can be summarized in the following way:

Glycogen $\Rightarrow$ glucose $\Rightarrow$ pyruvate $\Rightarrow$ reduction of pyruvate $\Rightarrow$ lactic acid $\Rightarrow$ only 3 molecules of ATP, but 38 molecules of lactic acid are obtained out of one molecule of glucose.

Formation of lactic acid during post mortem glycolysis is responsible for the decline in pH value within the meat after slaughtering. The pH value of meat at point of slaughter varies depending on the type of animal but is generally between 6.8 and 7.2. After slaughter, the pH normally drops to around 5.3–5.4, which is close to the isoelectric point of muscle meat. On completion of post mortem glycolysis, the final pH value of red meat is slightly higher than that of white meat: red meat generally contains slightly less glycogen than white meat at the point of slaughter, and as a result, the pH value does not decline as much during rigor mortis. High final pH levels in meat after completion of post mortem glycolysis and rigor mortis correlate with a low level of glycogen being present at point of slaughter in muscle tissue, leading to insufficient acidification of muscle tissue postslaughter and resulting in a final pH value around 6.2–6.4 (DFD meat).

2.2.2 ATP Levels Postslaughter and Rigor Mortis

In living muscle, the level of ATP is around 5 μmol/g of muscle tissue, and levels as low as 1.5 μmol/g are sufficient for contraction and relaxation of the muscle fibers. The level of ATP remains unchanged for a short time after slaughter as ATP is rebuilt anaerobically by CP binding to ADP to generate ATP. After a certain period of time postslaughter (in pork after about 2–4 hours, in chicken 1–2 hours, and in cattle 4–8 hours), the concentration of ATP in muscle tissue drops below 1 μmol/g of muscle tissue. At this point, meat exhibits a pH value of around 6.0 and Ca^{2+} ions are no longer absorbed by the sarcoplasmic reticulum. At such low levels of ATP, actin and myosin do not dissociate anymore and instead remain bound together to form the actomyosin complex. This point marks the onset of rigor mortis, and in rigor mortis, the head of the myosin filament "locks" into the actin and permanent "cross-links" are established between the two filaments by not separating any longer as no, or insufficient, ATP is provided. On completion of post mortem glycolysis, most actin and myosin filaments are cross-linked to form the actomyosin complex.

Post mortem glycolysis generally takes place slower in cattle compared with pigs. Certain breeds of pigs, such as Pietrain and German Landrace, tend to undergo very fast post mortem glycolysis and therefore are prone to developing meat with pale-soft-exudative character. Rigor mortis is complete in beef after around 24–36 hours and in pigs after around 12–16 hours. In poultry, such as chicken, it takes only 2–4 hours until rigor mortis is completed: chicken breast especially can demonstrate extremely fast rigor mortis. In extreme cases, rigor mortis is completed within 1–1.5 hours after slaughtering in chicken breast.

On completion of post mortem rigor mortis, the pH value has dropped to around 5.3 and actin and myosin are present as the actomyosin complex, contrary to the situation when the animal was still alive. Water-holding-capacity (WHC) and solubility of muscular protein are greatly reduced as a result of those changes. From this point on, meat enters the stage of maturing, also known as "ripening," and tenderization starts to take place given that the decline in pH value releases enzymes responsible for tenderness (Table 2.1).

TABLE 2.1 Differences in Meat Preslaughter and Postslaughter

	pH	WHC	Solubility of Protein
Preslaughter	6.8–7.2	Excellent	Excellent
Postslaughter	5.3–5.4	Poor	Poor

Chapter 3

Definitions

3.1 PSE/RSE/DFD AND "NORMAL" MEAT

Proper handling of cattle, pigs, and poultry before slaughter has a tremendous impact on the quality of meat obtained, and there is much research on and investment in abbatoir design and construction. Even the input of specialists such as animal psychiatrists is vital to establish a well-functioning and efficient slaughtering process. If the "functional quality" of the meat is destroyed during the slaughtering and cooling process, neither the use of an additive nor processing with modern equipment can remedy this damage.

Stunning with the use of carbon dioxide (CO_2) is the preferred method to knock out pigs, because the pigs lose consciousness gently and in comparison to electrical stunning, much less internal muscle bleeding occurs and fewer bones are broken. The pigs are placed in a cage and lowered into a chamber where the concentration of CO_2 gradually increases and remains at a high concentration for a certain period of time. Stunning pigs using a gas such as argon is an even more gentle way to knock out pigs, and an improved quality of meat could be obtained using this method. However, stunning pigs with the noble gas argon is in its infancy and further research and testing are required. It is usually more expensive to use argon than to use CO_2. To avoid acute "stress" before slaughtering, gentle showers, like a fine mist, are sprayed over the pigs to keep them "cool" and waiting areas are light blue or light green, which also calms down pigs. In addition, when pigs have to walk to the stunning area on their own feet, they prefer to walk slightly uphill, and they walk faster if they are walking toward a well-lit area, so abbatoirs often are designed with this in mind. Research continues to find better and more-efficient ways to improve both meat quality and slaughtering efficiency.

Selective breeding and especially overbreeding in pigs have produced significantly larger amounts of muscle meat in breeds such as Pietrain and German Landrace, and others, compared with 20 years ago. The aim of selective breeding is to maximize growth of the high-value cuts of a pig such the loin and the leg. Muscle tissue needs to be supported by a lot of oxygen, and the skeleton, as well as the size of the heart, has remained the same during the time that the amount of muscle tissue has increased dramatically. As a result, overbred animals are very sensitive to "stress" or stressful situations.

3.1.1 PSE/RSE/DFD Meat

Pale, soft and exudative (PSE) is the term used to describe a defective type of meat, seen predominantly in pork but also in poultry. Unfortunately, it is high-value cuts, such as loin and leg meat, that are predominantly affected, and it is the degree of PSE condition in meat that is crucial, as the borderline between PSE and non-PSE meat is not clearly defined.

The combination of two factors, a low pH value shortly after slaughtering and a high temperature within meat (above 37°C) at the same time, causes the PSE condition. In acute stressful situations before slaughter, pigs produce lactic acid from glycogen anaerobically while they are still alive and, they breathe heavily to form the large amounts of adenosine triphosphate (ATP) required. The formation of lactic acid, while the animal is still alive, in conjunction with fast post mortem glycolysis, causes a rapid drop in pH. This results in an abnormally low pH value in muscle tissue shortly after slaughter, contributing to the formation of PSE meat. Electrical stunning of pigs most commonly speeds up post mortem glycolysis and therefore also contributes to the formation of PSE meat. Stressful conditions before slaughter also cause a rise in pigs' temperature, and the temperature of meat remains high after slaughter. PSE-susceptible pigs can exhibit body temperatures of around 42°C if heavily stressed shortly before slaughter, whereas 37°C is the normal body temperature. A halothane test is frequently applied in order to determine PSE susceptibility in pigs. Halothane-positive pigs, commonly Pietrain and Landrace, demonstrate great tendency to produce PSE meat when the animal is not handled properly before slaughter.

PSE meat is usually pale in color, wet in appearance, and very soft in texture, and the PSE condition is caused by partially denatured proteins. Denatured proteins cannot hold, or bind, muscular water as well as fully native proteins. More specifically, the length of the myosin filament is reduced by around 8–10% during this process of denaturation, and water-holding capacity (WHC) of meat (the capacity of meat itself to retain, or hold, its own tissue water itself) is greatly reduced as a result. The reduced WHC explains the fact that PSE meat appears to be "wetter" than normal pork meat. However, the level of water within PSE pork is in most cases the same as in normal pork, but the WHC is reduced due to the smaller quantity of native protein in the meat. Firmness of PSE pork is also reduced compared with normal pork due to partial denaturation of proteins: denatured proteins exhibit a change in their three-dimensional structure, which leads to a less-firm structure overall.

The light, or lighter, color of PSE pork is explained by the small myofibrillar volume in the muscle tissues. Muscle tissue with a small myofibrillar volume has high light scattering ability, so light is reflected differently in PSE meat compared with normal pork. Light is unable to penetrate into the meat and so gets scattered right on the surface and the myoglobin cannot absorb the light, making meat appear pale. It is an interesting fact that the color of PSE pork is generally lighter even though the content of myoglobin in PSE pork is in most cases the same as that in normal pork. The small myofibrillar volume causes the open meat structure of PSE meat as denatured proteins shrink, resulting in larger gaps between the individual fibers. The L value (see Section 5.5 under Chapter 5: Color in Cured Meat Products and Fresh Meat) is greater than 50 in PSE meat.

To some degree, myoglobin is denatured as well in PSE meat, and denatured myoglobin does not contribute to the formation of curing color.

PSE pork can be checked 45 (pH_{45}) or 60 (pH_1) minutes after slaughtering by checking the pH value in the muscle of the loin (*Musculus longissimus dorsi*). If the pH_{45} is at 6.0 or the pH_1 is at 5.8, PSE meat is obtained. When pH_1 is below 5.8, severe PSE pork is the result. Checking the pH value after 24 hours, once the rigor mortis in pork is completed and the meat is well chilled, does not give any indication toward PSE given the fact that in PSE pork, as well as in normal pork, the final pH value is more or less the same.

Chilling pork carcasses fast, commonly in blast-chillers straight after slaughtering, helps to minimize the severity of PSE. If meat with a high pH value is chilled quickly after slaughter, the meat proteins are not damaged as much as they would be if the meat were chilled slowly. During such fast chilling, a temperature of 32–35°C is commonly reached within 90 minutes. Despite the possibility of obtaining a slight degree of cold shortening during "fast" chilling, a slight degree of cold shortening (see Section 3.1) is of less economical disadvantage compared with obtaining PSE meat.

Red, soft, and exudative (RSE) is another term used to describe a quality defect in meat. RSE pork has the same characteristics as PSE pork, except that it preserves the natural red color of meat better compared with PSE pork, possibly because the carcass was chilled quickly after slaughter and the L value is 44–50. Light is not scattered as severely as it is under "more severe" PSE conditions, and the meat appears redder in color despite the fact that a similar level of protein is denatured.

Neither PSE nor RSE meat offers technological advantages within the manufacture of meat products. Due to the reduced level of native proteins, WHC and water-binding capacity are reduced, and the "lighter" color is of no benefit. It is also important to emphasize that once proteins are denatured, as they are in PSE pork, no additive can make up for this shortcoming afterward during the production of meat products, and the nonfunctional proteins cannot be turned back into their native state. The level of PSE pork varies from country to country; some countries have as little as 2–4% PSE pork, while others obtain 25–40% of PSE calculated from the total number of pigs slaughtered (Fig. 3.1).

DFD is the abbreviation for dry, firm, and dark, and DFD meat is also known under "dark cutting beef" showing an L value of less than 44. DFD characteristics in meat can be seen predominantly in beef and lamb; however, some pigs also exhibit DFD character.

Contrary to pigs, which produce lactic acid from glycogen in an anaerobic way if exposed to "stress" before slaughter, animals such as cattle, deer, or lamb use glycogen in an aerobic way if stressed before slaughter. They simply burn

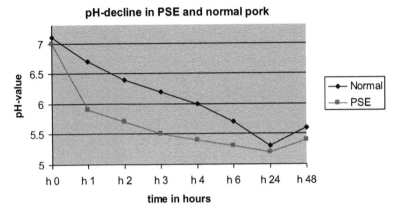

FIGURE 3.1 Decline in pH value of PSE meat in comparison with normal meat postslaughter.

energy (glycogen) under stressful situations for the formation of ATP, and no lactic acid is obtained while the animal is still alive. As a result, insignificant amounts of glycogen are left in the muscle at the point of slaughter and no, or very little, lactic acid can be produced postslaughter during rigor mortis. This results in an insufficient decline in pH value within meat after slaughter, and on completion of rigor mortis, by beef after around 24–36 hours, the pH value in meat is still around 6.0–6.2. This phenomenon is also sometimes called "incomplete rigor mortis" as proper and sufficient acidification within meat never takes place post mortem and a low number of cross-links between actin and myosin are established. The low number of cross-links explains the high solubility of DFD meat: the protein molecules are not as tightly bound together as they are in PSE meat or meat that has undergone a "normal" rigor mortis. DFD meat demonstrates a "closed" fiber structure once rigor mortis is completed, and only small gaps are present between the muscle fibers actin and myosin.

Contrary to PSE, where the pH value has to be checked 45 minutes or 1 hour after slaughter, DFD can be detected on completion of rigor mortis in beef after around 24–36 hours postslaughter. If at this point, the pH value is at (or above) 6.0, DFD meat is obtained. DFD meat appears dark in color due to the "closed" fiber structure (little gaps between actin and myosin) and has a slight slimy-tacky appearance, which is not microbiological sliminess caused by high numbers of bacteria. When cutting steaks of DFD meat, butchers use the phrase "the meat doesn't come off the knife" to describe their character.

From a technological point, DFD meat has the advantage of high protein solubility, as acidification during rigor mortis never really took place and an actomyosin complex was only obtained to a small degree. The WHC of DFD meat is also excellent: the high pH value correlates with a high WHC as the pH value of DFD meat is a long way from the IEP (pH value of 5.2). On the other hand, due to insufficient acidification of muscle tissue during rigor mortis, the shelf-life of DFD meat is dramatically shortened as nearly all types of bacteria

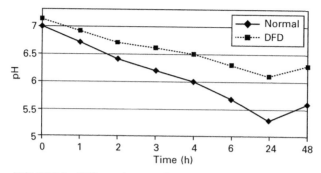

FIGURE 3.2 DFD meat in comparison to normal meat postslaughter.

TABLE 3.1 Differences in Certain Characteristics Within PSE or DFD Meat

Characteristic	PSE	DFD
pH-value	$pH_1 < 5.8$	$pH_{24} > 6.2$
WHC	Poor	Very good
Consistency of meat	Soft	Solid
Color of meat	Pale-light	Dark
Post mortem glycolysis	Very fast	Slow and incomplete
Shelf life of meat	Slightly reduced	Significantly reduced
Tenderness of meat	Reduced	Enhanced

find favorable conditions for growth at elevated pH levels. The high pH value is also an obstacle for the development of curing color in cured meat products produced from beef (see Section 5.3 under Chapter 5: Color in Cured Meat Products and Fresh Meat) (Fig. 3.2).

In poultry, both PSE and DFD character can be found. Poultry generally enter rigor mortis very quickly and post mortem glycolysis seems to take place faster in white muscles, such as breast, compared with red muscles from the leg (Table 3.1).

3.2 MECHANICALLY SEPARATED MEAT

Mechanically separated meat (MSM) is obtained from meat trimmings, which are high in connective tissue and cartilage such as shank meat. These materials are put through a machine that separates meat from connective tissue based on the different degree of firmness and texture of those materials. This process saves a lot of labor and, commonly, a "Baader" machine is used. The meat material obtained from this separation process is high in protein and an excellent material for all types of sausages and coarse minced salami.

The production of MSM takes place in the following stages. The material to be processed is commonly minced first with the 13- to 20-mm blade and subsequently fed into the "Baader machine." Lean meat, due to its soft texture, passes through holes in a rotating barrel into the inside of the barrel and is discharged through the open side on the barrel. Connective tissue and ligaments are discharged in front of the machine and do not penetrate into the inside of the barrel. This material is rich in connective tissue and can be perfectly used in cooked sausages as the high content of connective tissue and, therefore collagen, contributes very positively to the bite of a cooked sausage. Such collagen-rich material can also be applied in raw fermented salami (see Section 7.1.1 in Chapter 7: Fermented Salami: Non–Heat Treated, for salami). MSM contains around 15–17% protein: about 70–80% compared with protein in lean muscle meat (around 21% protein). Therefore, the water-binding and fat-emulsification capacity of soft MDM is around 70–80% of that of lean muscle meat, and all protein within MSM is still functional as it is not damaged during processing.

3.3 COLD SHORTENING

Cold shortening is the result of cooling down warm or hot carcass meat too quickly after slaughter. Cold shortening is predominantly seen in beef as well as lamb if the internal temperature of the meat reaches (or drops below) +14°C, while the pH value is still by around 6.0–6.2 at this stage during rigor mortis. A temperature of around +14°C is normally achieved only after 12–16 hours in cattle after slaughter, and in pork, a temperature below 15°C should not be obtained within 4–5 hours after slaughter.

Another such temperature–pH relationship is that temperatures of below 7°C within meat should not be obtained while the pH value is still around 5.8–6.0. The combination of high pH value and low temperature present at the same time in muscle tissue damages the sarcoplasmic reticulum (SR), and contraction, as well as relaxation, of the muscle fibers post mortem cannot be controlled properly anymore. When damaged in this way, the SR does not reabsorb the Ca^{2+} ions released for contraction, and permanent high concentrations of Ca^{2+} ions, as well as the nonactivation of the enzyme actin-myosin ATPase due to the damaged SR, cause the muscle fibers to contract heavily. All energy obtained in an anaerobically postslaughter is used for muscle contraction only, and a large number of cross-links between actin and myosin are established. As a result, meat is always tough and the solubility of the protein is greatly reduced as well, as solubility correlates with the numbers of cross-links within muscle tissue.

Cold shortening does not occur at, or below, a pH value of 6.0, and once such a pH value is obtained, the carcass can be chilled more rapidly but, as said earlier, temperatures in muscle tissue below 7°C at this point must be avoided. Another way to avoid cold shortening is the 10/10 rule, meaning that the temperature within meat on a carcass should not be below 10°C within 10 hours after slaughter. This rule of thumb applies to carcass meat with bone-in. When meat is deboned in a hot stage of processing, the boneless meat should not be chilled below 16°C within 10 hours. The bone structure of carcass meat counteracts the contraction of the muscle fibers, and given that this counterforce is not in place by deboned meat, chilling of boneless meat has to take place at a slower rate.

Red muscles are more susceptible to cold shortening compared with white muscles given that white muscles demonstrate a more sophisticated and developed SR, which is responsible for the release of Ca^{2+} ions. The SR of white meat can reabsorb Ca^{2+} ions more effectively compared with red meat, and the impact from the damaged SR in white meat is not as strong as in red meat. The release of Ca^{2+} ions post mortem in high amounts results in a strong stimulation of the fibers actin and myosin, which subsequently leads to strong contraction. If those ions are not reabsorbed effectively, as it takes place in red meat, toughness is the consequence.

Contrary to cold shortening, rigor shortening occurs when the carcass is cooled down too slowly. The combination of a pH value around 5.8–6.0 shortly before the onset of rigor mortis with high temperatures in muscle tissue above 22°C causes severe shortening known as rigor shortening; as a result, the muscle fibers can shorten up to 25% from their original length. In extreme cases, when the temperature of the meat is around 30°C at mentioned pH levels, shortening of the fibers can be up to 45% from its original length. An associated problem with cooling down carcasses too slowly is that a high temperature is found in the large, or thick, portions of meat for a prolonged period. Enzymes present in such warm areas become very active, and decarboxylation (splitting away of a COOH group) occurs, which can give the meat a rusty-green color as well as an off-putting smell. During this process, large amounts of acetic and butyric acid and toxic biogenic amines are formed.

Cold shortening can be observed occasionally in pork and predominantly in such muscles where rapid chilling took place on an exposed muscle surface. In pork, a slight degree of cold shortening is sometimes desired in order to reduce the level of PSE meat (see Section 3.1) given that a slight degree of cold shortening is less damaging economically than obtaining PSE meat.

Very fast chilling postslaughter at temperatures of around −20°C without electrical stimulation does not result in cold shortening, but this technique is not widely applied yet. This could be because the application of such low temperatures, in conjunction with high air velocity, to the surface of the meat results in "hardening", or even slightly frozen conditions, on the surface of meat. Despite the continuous impact of low temperatures, the degree of shortening is reduced, as the hardened fibers just do not contract severely anymore.

The effect of gravity on the carcass is also significantly reduced when muscles are hardened. However, in general it is only possible to attain a temperature of around 0°C within 4–5 hours after slaughter within thin pieces of meat while not dropping the temperature below −1.2°C in order to avoid freezing. The important parameter for this technique to avoid cold shortening by very fast chilling is the use of thin pieces of meat and not whole carcasses as meat (containing a lot of water) generally is a very poor conductor of heat. It is practically impossible to lower the temperature within thick pieces of carcass meat to 0°C in the core quickly while avoiding freezing of the meat material on the surface as would be caused by dropping the temperature below −1.2°C.

3.4 FREEZING AND TEMPERING OF MEAT

Worldwide, a huge amount of meat is handled and distributed in a frozen state as storage of meat over a prolonged period of time can only be achieved by freezing. Within the production of fermented salami, fat material is generally processed fully frozen, while meat or meat trimmings are most often processed in a tempered state. Bacterial growth generally stops at around −12°C, but bacteria are not killed at such temperatures and it is vital to understand that freezing does not improve the microbiological status of meat. Some strains of mold are known to even grow at temperatures as low as −14°C and below. Any meat or fat to be frozen should be packed or wrapped in some kind of foil or bag in order to avoid freezer burn (see Section 3.5). Packing of meat also avoids dehydration on the surface of the frozen material and acts as a form of protection against contamination during handling. Freezing is a highly energy-consuming process, and therefore in general it is deboned meat that is frozen, eliminating the cost of freezing bones. The conversion of 0°C water into 0°C ice requires the extraction of around 300 kJ of "latent heat" per kg of water. In comparison, the cooling of water from 30 to 0°C requires only around 100 kJ/kg of water. Nevertheless, certain cuts such as pork legs or whole chicken are commonly frozen without being deboned.

During freezing, cellular water is turned to ice by removal of energy from the water. In the first stage of freezing, the decline in temperature within chilled meat from +4 to 0°C is very fast. Reducing the temperature further from 0 to −10°C takes a long time, and a large amount of latent energy must be removed in comparison to the previous step. As explained earlier, turning water into ice is a highly energy-consuming process, and in particular, the change in phase from water into ice uses a lot of energy. The subsequent step to reduce the temperature even further from −10 to −18°C occurs again quickly once the majority of water is already present as ice (Fig. 3.3).

The freezing point of water within muscle tissue is around −1.2°C due to the natural presence of salt in meat, and this lowers the point at which water turns into ice. Even at temperatures as low as −18°C, not all water is frozen and around 1–1.5% remains as water within the cell. By −3°C, around 40% of the total water in a cell remains as water, and freezing of all water in meat requires a temperature of around −49°C.

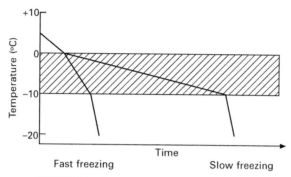
FIGURE 3.3 Temperature curve by freezing meat.

Different speeds of freezing are as follows:

- Slow freezing: <0.5 cm/h
- Fast freezing: >0.5 cm/h
- Very fast freezing: <10 cm/h (vegetables)
- Ultrafast freezing: >10 cm/h (tomatoes, cucumbers)

Freezing of meat or fat by exposing these materials to temperatures of around −18 to −25°C is a slow freezing process, and meat is often frozen to temperatures within this range. Fast freezing using cryogenic gases such as CO_2 or nitrogen, though, is regularly used in the production of burgers or crumbed portioned food, such as chicken nuggets.

Large ice crystals are formed in the extracellular space during slow freezing, and these crystals grow further in size as water from the intracellular space is attracted to them. Water is frozen out of the extracellular space (ie, it separates from other substances), and therefore the concentration of salt within the extracellular space increases. Due to the imbalance in salt concentration, water diffuses from the intracellular space toward the extracellular space in order to reach an equilibrium, and the cell is damaged chemically as the concentration of salt in the extracellular space increases. The cell is also damaged physically both by the water freezing out and by the low water content in the extracellular space. Last, the large ice crystals formed damage the cell mechanically, by applying pressure to cell membranes. The crystallization of ice occurs in three steps. First, tiny ice crystals are formed once water has turned into ice at a slow freezing speed. Second, the tiny ice crystals increase in size as liquid water is attracted to ice crystals, and last, a network of ice crystals, made of large ice crystals, is formed (Fig. 3.4).

Several different methods of freezing are used in the meat industry. Plate freezers work on a semicontinuous basis, and the material to be frozen is exposed to a large surface area on both sides. Plate freezers commonly operate with liquid ammonia, and the temperature on the surfaces of the plate freezer is around −35 to −40°C. Despite that, freezing using plate freezers is still categorized as slow freezing. Materials frozen by the help of plate freezers are

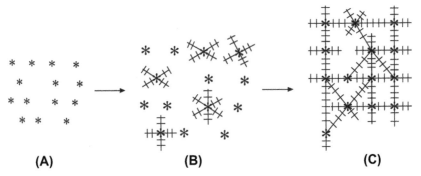

FIGURE 3.4 Formation of ice crystals: (A) small isolated crystals; (B) larger cross-linked ice crystals; (C) large connected ice crystals.

generally of block or rectangular shape. Blast freezers operate by forced circulation of air and function at temperatures of around −35 to −40°C with an air velocity of around 4–6 m/s. Generally speaking, blast-freezers operate at around −20 to −25°C, but the high air velocity reduces the temperature to the abovementioned range of around −35 to −40°C, with blast freezing still being considered a slow freezing process. The storage time of meat and fat materials under freezing conditions is largely determined by the amount of fat present within those materials as well as by the amount of unsaturated fatty acids within the fat itself. As a general rule, an increased fat content shortens storage time given the fact that rancidity develops at a faster rate within fat or fatty materials.

Lean beef can be stored for about 1 year, lean veal for about 9 months, lean pork for about 6 months, and pork fat for about 3–4 months at a temperature of around −18 to −20°C. The difference in storage time between lean beef and lean pork is due to the fact that the fat within pork contains a significantly higher level of unsaturated fatty acids compared with beef fat and is therefore more susceptible to rancidity. Maintaining a constant temperature during storage under freezing conditions is vital to avoid any processes of diffusion or enzyme activity, which could lead to a greater degree of rancidity.

3.4.1 Tempering of Meat

Tempering (and thawing) is the reverse process of freezing and is required for smooth processing during cutting and mincing processes during the production of salami. The major difficulty when tempering is that the difference in temperature between the source of heat and the frozen meat cannot be too great for a prolonged period of time as this causes bacteria to grow at a fast rate, creating a microbiological risk.

Tempering takes place at a slower rate than freezing as water displays less thermal conductivity than ice. During freezing, the layer of ice formed on the outside of the material to be frozen removes energy quickly from the water still present within the inner layers of the material and, as a result, the water turns fairly quickly into ice.

During tempering, though, water is present first on the surface on the outer layers of the material to be defrosted. As water is a poor conductor of heat, the heat penetrates very slowly through the layer of surface water first before finally reaching the ice in the inner layers and core. Tempering of meat for salami takes usually place by placing blocks of frozen meat on racks in the chiller at temperatures between +2 and +5°C for around 2–3 days before being further processed. A less time-consuming process of tempering is to pass fully frozen blocks of meat through an industrial-sized microwave oven, raising the temperature from around −20°C to around −6 to −8°C within a few minutes before meat is further processed.

3.5 FREEZER BURNING

Freezer burning on frozen meat occurs if meat is stored unpacked under freezing conditions. Due to the circulation of air in a freezer, ice present in the outer layers of frozen meat sublimes to gas. Sublimation is a process where water turns from its solid state (ice) into its gas-like state without ever being present in its liquid state (water). As a result of sublimation, the macromolecules within the outer layers of frozen meat change their configuration and proteins are denatured during the process. Following sublimation, less functional, or native, protein is available during processing and the head of the myosin molecule is predominantly denatured, although other parts of the myosin molecule, as well as actin, seem not to be affected much. Denaturation of proteins is also the result of a high concentration of salt within the outer layers, given that the level of moisture is low in comparison to salt present naturally in meat. The combination of a low water activity (Aw), in conjunction with an increased level of salt, partly denatures protein.

Meat that experienced freezer burning exhibits on its outside layers a dry and fibery structure due to severe dehydration. Changes in color within those layers can be observed as well, and the original red color changes into a lighter, sometimes even slight yellow/green color. Rancidity is also sped up in those dry outside layers as a reduced water content favors the development of rancidity. Freezer burning can be largely avoided if meat is frozen in a packed form. When packaged, water cannot sublime and moisture is not lost when ice turns into gas (sublimation) due to air circulation in the freezer. If the product is packed or covered properly, the packaging material does not allow the gas to evaporate. The packaging material should be of low water permeability, and as little space as possible should be present between meat and the packaging material due to possible oxidation in those areas.

3.6 pH VALUE

The pH value is of crucial importance during the manufacture of acidified and dried salami as several major processes regarding food safety, sliceability, curing color, and texture are closely related to the pH value within a salami. The abbreviation "pH" stands for the Latin term *potentia hydrogeni*, or "potential of hydrogen" (effectiveness of hydrogen). The pH value has a significant impact on color, shelf-life, taste, microbiological stability, yield, and texture of meat and meat products and is therefore one of the most important parameters within the production of meat products and meat itself. Often, the pH value is referred to as "the acidity of meat," which is just partly correct given that the pH scale ranges from 0 to 14 and does not the cover only the sour range. The pH value of meat and meat products lies generally between 4.6 (raw fermented salami) and 6.4. At a pH value of around 6.4, meat is spoiled due to enzyme activity, which produces a large amount of metabolic byproducts as well as ammonia. Sliminess, bad smell, and discoloration can be seen at this point as well (Fig. 3.5).

A pH value of 0.1 is extremely acidic, 7 is neutral, and 14 is extremely alkaline.

pH > 7—alkaline solution
pH = 7—neutral solution
pH < 7—acid solution

Mathematically, the pH value is the negative logarithm to the base of 10 of the hydrogen ion [H^+] concentration ($-\log_{10}$ [H^+]). The pH value expresses the concentration of [H^+] in molarity, or moles per liter, but not in percent. A change of 1 pH unit indicates a 10-fold change in the concentration of hydrogen ions. The increments of pH are logarithmic, not linear. A shift in pH value from 6 to 5 represents 90 units of hydrogen ion activity, whereby a shift from 5 to 4 represents 900 units.

In a neutral solution, the concentration of [H^+] and [OH^-] ions is equal, or of the same number. In such an example, the concentration is 10^{-7} mol H^+/L (or 0.0000001 mol H^+/L). Taking the negative logarithmic of such a concentration (10^{-7}) of [H^+], the number 7 is obtained. Therefore, the pH value of this particular solution is 7, which is neutral and this particular example represents the concentration of [H^+] in pure water at 25°C.

If the concentration of [H^+] is 0.00001 mol H^+/L of water within meat, or 10^{-5}, the pH value of such a solution is 5, and therefore acid. On the other hand, if the concentration of [H^+] would be 0.01 mol H^+/L, or 10^{-2}, the pH value would be 2, which is very sour. The basic principle behind this is that a lower exponent results in a higher degree of acidity and a 10-fold increase in [H^+] corresponds to a decrease of 1.0 pH unit. Because of the logarithmic nature and

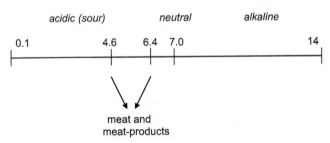

FIGURE 3.5   Scale of pH value.

because it is based on the amount of [H⁺] ions present, the pH value decreases if the concentration of [H⁺] ions increases.

3.7 Aw VALUE

Similar to the pH value, the Aw value also plays a significant part in the production of fermented and dried salami. The Aw value is the ratio of the vapor pressure above a solution compared with the vapor pressure of pure water at the same temperature. In physical terms, the Aw can be calculated in the following way:

$$Aw = P/P_0$$

where P, vapor pressure of food at a certain temperature in °C and P_0, vapor pressure of pure water at the same temperature in °C.

Another (simple) way of expressing Aw is the relative humidity (RH) of food divided by 100 (for example, 98.5/100 results in an Aw of 0.985).

It is often a point of confusion that the Aw value in meat or meat products is not the total water content but rather the amount of free unbound water within the product in relation to its total water. To be more specific, pure water, without any impurities and no minerals in it, demonstrates an Aw of 1.00, and when discussing Aw value, at least two figures after the comma have to be used. By an Aw of 1.00, all water is unbound ("free") and available for bacteria as a source of food. Fresh meat exhibits an Aw of around 0.98, meaning that around 98% of the total water within meat is unbound, while at an Aw of 0.80, significantly less free water is present in meat. Air-dried cured meat products commonly display an Aw between 0.82 to 0.90, while spreadable raw sausage frequently shows an Aw of 0.94. Hungarian salami normally exhibits an Aw of around 0.83–0.85 and the Aw is largely determined by dried meat products by the degree of drying itself as well as the concentration of Aw-reducing substances such as salt and sugar. On the other hand, cooked sausages, as well as cooked ham, regularly exhibit an Aw of around 0.97–0.98 due to that fact that more or less water is added to the product during processing.

Well-dried salami, after losing around 30% in weight during fermentation and drying, exhibits an Aw of 0.90–0.88. Despite the fact that an Aw of 0.90 still seems high, at such levels raw fermented salami is shelf stable without refrigeration given that harmful microorganisms such as *Staphylococcus aureus* and *Salmonella* spp. do not produce toxin at those Aw values. This simple example highlights why two figures after the decimal point have to be used because by only going from an Aw of 1.00 (free water) or fresh meat (0.98) down to 0.90 (shelf stable salami), dramatic changes in physical behavior, appearance, and shelf-life of a product can be seen. The Aw value in meat products can be reduced by the impact of drying (removal of water), the addition of sugar or salt to food (binding of water through ions), the addition of fat (less water added to the meat product), or freezing (immobilization of water).

3.8 Eh VALUE (REDOX POTENTIAL)

Redox (reduction–oxidation) reactions are one of the most commonly occurring reactions in cells and are mostly catalyzed by enzymes. Reduction and oxidation refer to the exchange of electrons from a donor to an acceptor where the donor is oxidized and the acceptor is reduced at the same time. A molecule taking up (gaining) electrons is the oxidizing agent (oxidant), and by gaining an electron, the oxidizing agent is reduced. The reducing agent (reductant) gives away, or donates, electrons and is oxidized as a result. The process of oxidation is a releasing, or removal, of electrons, while the process of reduction is a gaining, or addition, of electrons. According to the law of electroneutrality, the oxidation of one substance always corresponds to the reduction of another substance. An oxidizing agent changes via a reduction (gain of electrons) to a reducing agent. A reducing agent turns via an oxidation (donation of electrons) into an oxidizing agent (Fig. 3.6).

The most well-known example for such a redox reaction couple is the reaction of Fe^{2+} with oxygen in the hemoglobin system.

$$Fe^{2+} \Rightarrow Fe^{3+} + e^{-} \text{ (oxidation)}$$

$$O_2 + e^{-} \Rightarrow O_2^{-} \text{ (reduction)}$$

$$Fe^{2+} + O_2 \Leftrightarrow Fe^{3+} + O_2^{-} \text{ (redox} - \text{reaction)}$$

In this example, twice positively charged iron is oxidized by the impact of oxygen. Since Fe^{2+} gives away an electron to oxygen, it is a process of oxidation. Oxygen, taking up the electron, is the oxidizing agent, while Fe^{2+} is the reducing agent due to the donation, or giving away, of an electron.

Both substances act as the redox couple, and such oxidation–reduction reactions cause the exchange of free energy. The flow of electrons in the exchange of free energy can be measured and is called the redox potential or electromotive force, a measurement for oxidizing or reducing substances expressing the ability of absorbing or donating electrons (electron affinity) within food. The redox value is expressed in units of millivolt (mV) and is the amount of energy gained by transferring 1 mol of electrons from an oxidant to H_2.

The Eh value depends largely on the chemical composition and the oxygen pressure within food as well as the pH value. A reduced Eh value lowers the ability for aerobic bacteria to grow, while a high Eh value reduces the growth of anaerobic bacteria. A lowering of Eh value can be achieved by the addition of reducing agents such as ascorbic acid/erythorbate or the application of vacuum during the manufacture of a product.

$$\text{Reducing agent} \underset{\text{Reduction}}{\overset{\text{Oxidation}}{\rightleftarrows}} \text{Oxidizing agent} + \text{Electron}$$

FIGURE 3.6 Oxidation–reduction.

3.9 CONDENSATION WATER

Formation of condensation water is frequently seen on fermented salami on the start of fermentation as well as, although not desired, during slicing of products. Air can hold (bind) moisture according to its temperature, and warmer air can hold more moisture compared with cold air. The term "relative humidity" (RH) refers to the actual level of moisture in the air compared with the moisture content of saturated air at the same temperature. There is a substantial difference regarding the amount of moisture in air between 75% RH at +4°C (as in a meat chiller) and 75% RH at 15°C. The amount of moisture in the air at +4°C is significantly less compared with +15°C.

Formation of condensation water on meat and meat products is frequently neglected and leads to microbiological spoilage, discoloration, and a significantly shorter shelf-life. Condensation water is formed on the surface of a cold material if such cold materials are exposed to a much warmer environment. When the cold material comes into contact with the warmer environment, the dewpoint is exceeded, and the amount of moisture, which cannot be bound by the air surrounding the cold material, can be seen as free water on the surface of the cold material.

A peeled and well-chilled salami to be sliced and packed, for example, coming out from the chiller (fridge) exhibits a surface temperature of +3°C and the temperature in the slicing room is +12°C. A cold microclimate, surrounding the cold salami, forms and the temperature within this microclimate is more or less the same as the salami itself, +3°C. The ability of this microclimate (+3°C) to hold moisture is significantly less than the ability of the significantly warmer air present in the slicing room (+12°C). As a result, the amount of moisture, which cannot be bound by the air within the cold microclimate "falls out" as the dewpoint is exceeded and can be seen as free water on the surface of the salami. This condensation water is, as said, free unbound water, and it is fully accessible for all microbes to use as food. Fig. 3.7 illustrates the described example.

In order to not exceed the dewpoint and therefore not obtaining condensation water on the surface of the product, the maximum room temperature

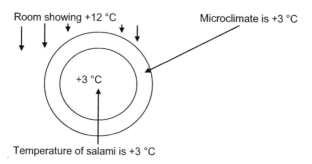

FIGURE 3.7 Formation of condensation water.

TABLE 3.2 Temperatures of the Meat Product and Room Temperatures at Various Levels of RH

Surface Temperature of Meat Product (°C)	Maximum Temperature (∝) in the Room at the Following RHs			
	60%	65%	70%	75%
0	7.5	6.5	5	4
1	8.4	7.6	6	5
2	9.5	8.5	7	6
3	10.4	9.5	8	7.2
4	11.8	10.6	9.5	8.5
5[a]	13.0	11.5	10.4	9.7
6[a]	14.1	12.7	11.5	10.4
7[a]	15.5	13.8	12.4	11.4
8[a]	16.4	14.9	13.7	12.7

[a]Temperature that should never be present for meat or meat products. They are cited only for the purpose of completing the table.

in this case would be +7.2°C if the room has an RH of 75%. To avoid the formation of condensation water on meat and meat products, the following can be done:

1. Reducing the temperature in the room, where meat or meat products are placed, in order not to exceed the dewpoint based on the difference in surface temperature of the product and air within the room.
2. Reducing the RH in the room by not changing the temperature in the room.

Table 3.2 shows the maximum temperature in the slicing or packaging room in °C in relation to the temperature of the meat or meat products in order to avoid formation of condensation water at a certain level of RH in the room:

(All temperatures marked with an asterisk [*] should never be present on meat or meat products and are mentioned for completion purposes only.)

3.10 MAILLARD REACTION

The Maillard reaction, named after L. C. Maillard, is also known as nonenzymatic browning. It is an extremely complex process and is the reaction between reducing sugars and proteins by the impact of heat. The Maillard reaction starts with a reducing sugar reacting with an amine, creating glycosyl amine. These substances undergo a reaction called Amadori rearrangement to produce a derivate of amino deoxy fructose. The reaction is continuous, and very reactive intermediate substances are formed, which subsequently react in several different ways. Eventually, a furan derivate is gained, and this derivate reacts with other components to polymerize into a dark-colored, insoluble material containing nitrogen. The Maillard reaction also takes place at room temperature but at a much slower rate and occurs at its slowest by low temperatures, low pH, and low Aw levels.

Excellent examples of the Maillard reaction are the crust of roast pork or browning of salami on pizza. The Maillard reaction also creates, besides color, countless complex flavors at the same time when, for example, salami is placed on a pizza and baked under high heat. The sulfur-containing amino acids methionine and cysteine play a primary role within the formation of the flavor-intensive components gained during the Maillard reaction. The flavor of such roasted or baked (application of hot air) meat products also contains heterocyclic compounds derived from amino acids, nucleotides, and sugars from the Maillard reaction such as oxopropanol and hydroxy-methylfurfural. Unsaturated fatty acids, as well as aldehyde fatty acid components, also contribute to the formation of odorous heterocyclic flavor compounds during the Maillard reaction.

Part II

Additives

Chapter 4

Additives

Additives are materials applied during the manufacture of food to increase, restore, or enhance attributes such as taste, color, texture, firmness, and shelf-life. Although the "dangers" of food additives are commented on almost daily in the media, food additives are in fact one of the most researched substances in the world. Before an additive is permitted to be applied to meat products, it must fulfill three parameters:

1. It must be technologically needed.
2. It must not be a threat to the human health.
3. It must not mislead the consumer.

Most countries today try to limit the number of additives permitted in meat products. In a situation where there were, for example, five permitted "emulsifiers," an additional sixth one would probably not be authorized unless there was a serious technological need. In a case where there is a need for a "new" additive, another hurdle is to prove that consumption of this additive over a certain period of time (mostly years) at a certain level does not have any negative impact on human health. This prerequisite is commonly fulfilled by conducting long-term-studies based on a certain ADI (available daily intake) level. The additive must also not disguise a quality defect in a food, caused by poor manufacturing practice, or hide a microbiological risk. An additive to improve the appearance of discolored meat, for example, would not be permitted.

4.1 PHOSPHATES

Phosphates are generally not applied in the manufacture of acidified and dried but do find an application in cooked salami. Phosphates are the salts of phosphoric acid and are widely applied in the meat industry. These salts are made out of positively charged metal ions and negatively charged phosphate ions, which are derived from the corresponding acid by loss of H^+. For example, sodium tripolyphosphate (STPP) falls apart in solution into Na^+ and PO_4^- ions and the negatively charged PO_4^- ion originates from the phosphoric acid in first place.

4.1.1 Production and Properties of Phosphates

Phosphates can be made in several ways but the most common way is as follows:

1. Phosphoric acid is obtained from raw or rock phosphorous, commonly with the help of other chemicals.
2. The phosphoric acid is cleaned in several stages to obtain food-grade phosphoric acid.
3. The food-grade acid is neutralized by mixing with a strong alkaline. Monophosphates (a salt) are obtained as a result of a reaction between the strong acid and alkaline, and water is removed during the process. Monophosphates show one phosphorous atom within the molecule.

The next steps involve the manufacture of higher-polymer phosphates, which takes place via a condensation-reaction at temperatures around 600–800°C by melting individual monophosphates together to form di-phosphates, also known as pyrophosphates. Pyrophosphates contain two phosphorous atoms in the molecule. Once another monophosphate is added on, tripolyphosphate is obtained and so forth. Within such a condensation-reaction, polyphosphates showing up to 1000 phosphorous atoms in the molecule can be produced.

4.1.2 Properties of Different Phosphates

There are three basic forms of phosphates:

- Ring—phosphates
- Chain—phosphates (linear phosphates)
- Combinations of ring and chain phosphates

In most countries, only chain phosphates (linear) are permitted to be applied in the meat-processing industry and in the treatment of seafood.

Phosphates fulfill several functions in meat products:

1. They "neutralize" the cross-link between actin and myosin, formed during rigor mortis, and support the dissociation of the actomyosin complex into separate fibers again. Phosphates loosen the electrostatic forces within the actomyosin complex: this function of phosphates is known as the "specific effect" on the muscular protein, as it contributes greatly to the solubility of muscular protein. Only phosphates are able to separate actin and myosin after rigor mortis and that is the primary reason for the worldwide use of phosphates. The separation of actin and myosin takes place as a result of the binding of the negatively charged phosphate ions with the positively charged Mg^{2+} or Ca^{2+} ions. The positively charged Mg^{2+} and Ca^{2+} ions play a vital role in muscle contraction as well as relaxation and are present at the point where binding between actin and myosin occurs by myosin locking into actin.

2. Through the addition of salt as well as phosphates at the same time to a meat product, the muscular protein becomes soluble and solubilized, or activated. The protein can then immobilize high levels of added water as well as emulsify a large amount of fat, given that activated meat protein is an excellent emulsifier of fat.
3. Nearly all phosphates, as well as blends of phosphates, used in the meat-processing industry are alkaline phosphates, and the addition of alkaline phosphates to slightly sour meat leads to a rise in pH inside the meat product. A movement further away from the iso-electric-point (IEP) in meat (pH around 5.3) takes place and enhanced water-binding-capacity of the protein is the result given that greater electrostatic repulsive forces create larger gaps between actin and myosin and larger amounts of added water can be bound.
4. The addition of phosphate, which is a salt itself, increases the ionic strength of the meat and an increased ionic strength leads to a more severe degree of swelling from the muscle fibers and activation of protein. Enhanced levels of activated protein support the immobilization of added water to meat products and the emulsification of fat.
5. Phosphates are slightly bacteriostatic and growth of bacteria is marginally slowed down. This slow in growth, though, is almost negligible in meat products as the concentration of phosphates required to show a significant impact regarding bacteria growth would be greatly above the permitted level.
6. Phosphates can chelate (bind) heavy metal ions and therefore slow down the process of rancidity as heavy metal ions are pro-oxidative materials.

In meat-related applications such as cooked salami, blends of different phosphates are commonly introduced as such blends are almost tailor-made for a specific application and perform significantly better compared with a single type of phosphate. A phosphate-blend used for the production of cooked salami contains predominantly the short-chain pyrophosphates and tripolyphosphates as those are the preferred phosphates regarding emulsification of fat and immobilizing added water. Short-chain phosphates, because those phosphates act on the protein right away, as required in such an application, and the most active component on the muscular protein is the pyrophosphate. Phosphate blends for emulsified sausages exhibit most often a pH value between 7.0 and 8.3 (Fig. 4.1).

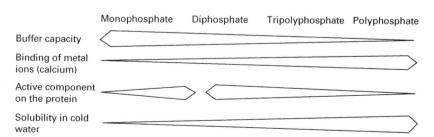

FIGURE 4.1 Properties of different phosphates.

Phosphates themselves do not activate much protein at all, but their function is to remove the links out of the actomyosin complex. Once the links are removed, the addition of salt and water causes the protein fibers to swell, and activated, or solubilized, protein is ultimately obtained. Longer-chain phosphates in sausages such as cooked salami and frankfurters help to increase the emulsification of fat and are present in phosphate blends for recipes that use a low meat content and at the same time high levels of fat and water. This type of phosphate blend frequently has a pH value of around 8.5.

There are also liquid blends of phosphates on the market, which generally contain around 20–25% P_2O_5 equals around 35–40% phosphates. So that the high percentage of phosphates in these blends dissolves in water, potassium phosphates are primarily used. The concentration of P_2O_5, though, cannot be more than around 27%. If higher levels of P_2O_5 were present in the solution and the solution were stored under cold conditions, phosphate crystals would form and never redissolve.

4.1.3 P_2O_5 Content of Phosphates

The P_2O_5 content represents the pentoxide content of a phosphate and is expressed in percent. It is a figure that expresses the actual phosphate content of a phosphate without the mineral part such as sodium, potassium, or calcium attached to it. As an example, STPP shows a P_2O_5 content of around 58%, meaning that of the total molecule, around 58% is the phosphate portion and sodium accounts for 42%. Even though blends of different phosphates are preferably applied in meat applications, the P_2O_5 content of a phosphate blend does not give any information about the functionality of the blend. As the P_2O_5 content shows the overall percentage pentoxide content of a blend of different phosphates, it does not reveal the percentage of each type of phosphate present within the blend. Only the proper combination of different phosphates accounts for the functionality inside the meat product afterward, and this is not expressed by the P_2O_5 content.

4.2 SALT—SODIUM CHLORIDE

Salt, or sodium chloride (NaCl) by its chemical name, is the world's oldest food additive and generally the most important additive in the production of meat products. The level of salt added when producing fermented dried salami is between 2.6% and 2.9% and as such significantly higher compared with most other meat products. This high level is based on the fact that salt acts as a preservative right from the start when producing salami, counteracting the risk of microbiological spoilage. By percentage, salt consists of 39.3% sodium and 60.7% chloride and frequently contains a small amount of anticaking agent to keep it free-flowing over a period of time. Most often trace elements are also present at around 0.2–0.3%. Salt is produced my mining (rock salt) or

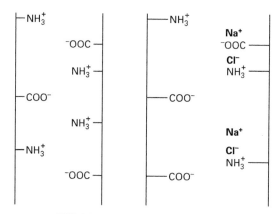

FIGURE 4.2 Impact of salt on protein.

evaporation (sea salt). Sodium is an essential nutrient and a material, which the body cannot produce by itself. Insufficient supply of sodium is a threat to the nervous and muscular systems within the human body, while oversupply of sodium leads to negative effects such as high blood pressure. In solution, NaCL hydrolyzes into Na^+ and Cl^- ions. The addition of salt to meat enhances the ionic strength and the Na^+ and Cl^- ions of salt bind to the ions of the side chains within the proteins and act as separating force between the side chains (Fig. 4.2).

Chloride is important as it aids potassium absorption in the human body. It is a component of the digestive stomach acid and enhances the ability of blood to carry carbon dioxide from respiring tissues to the lungs. The salty taste of a meat products containing sodium chloride originates predominantly from the positively charged Na^+ ions and to a small degree from the negatively charged Cl^- ions. In sodium-reduced meat products, potassium chloride is frequently the material of choice and potassium produces both a salty and often an undesired bitter taste. The maximum amount of NaCl that can be dissolved in 100 g of 20°C cold water is 35.8 g (a 26.4% salt solution). At this level, the solution becomes saturated as insufficient free water is available to dissolve any further salt. A 26.4% salt solution boils at 109°C. In frozen food, salt acts as a pro-oxidative and is commonly applied in frozen ready-to-eat meals in an encapsulated form.

Salt fulfills several functions in meat and meat products:

- Salt is a flavor enhancer, and no meat dish or meat product tastes good with insufficient salt, even if spices are used in the preparation of the meat.
- Salt, in conjunction with phosphates, solubilizes protein, which in turn can immobilize large amounts of added water and is able to emulsify fat in meat products. The addition of salt influences the interactions between actin and myosin. These electrostatic interactions are based on negative and positive

charges, which attract or repel each other, and the addition of salt causes a repelling effect, obtaining larger gaps between actin and myosin. Around 12 g of added salt per 1 kg of meat product is the lower limit to effectively activate protein.
- The texture of meat products is also improved by activation of protein.
- Salt lowers the Aw value (lowers the amount of free water within a product). Therefore, in meat products, such as raw fermented salami, it is an important hurdle against microbiological spoilage during the initial stages of the production.
- The addition of salt favors the growth conditions for gram-positive bacteria instead of gram-negative bacteria. Many human pathogens, such as *Salmonella* and *Escherichia coli*, are gram-negative bacteria.
- Salt itself eventually becomes poisonous to bacteria by creating an electrolyte imbalance within the cell.
- The addition of salt to meat causes a slight move from the IEP of muscle tissue toward a more acidic pH value. Depending on the amount of salt added, the IEP can move from 5.2 to around 5.0.

As a result, increased levels of water can be bound without changing the pH value of the meat itself, as the shift of the IEP from 5.2 to 5.0 widens the gap between pH value present in the meat and the IEP. For example, prior to the addition of salt, the pH gap in meat was 0.5 pH unit (5.7–5.2), and after the addition of salt, the gap is 0.7 pH unit (5.7–5.0).

A larger gap between the two pH values increases the capillary effect of the muscle fibers, and an increased capillary effect causes increased water-binding capability once again.

Salt, or specifically the sodium part of salt, can lead to high blood pressure if consumed in excess. "Light" meat products are available that have a sodium level of around 450–750 ppm of sodium per 100 g of product (depending on the food legislation in the respective country). In such products, only around 8 g of salt (sodium chloride) is applied per 1 kg of product, and potassium chloride is introduced instead so that the total level is 12–16 g of salt per 1 kg of product. When considering the sodium level of a meat product, it should also not be forgotten that sodium is frequently added to meat products in other forms, such as sodium nitrite, sodium erythorbate, and sodium phosphates. However, as explained earlier, a reduction is salt (sodium) levels within salami presents a food safety risk because the elevated level of salt within a salami acts as a preservative, supporting food safety.

4.3 POTASSIUM CHLORIDE AND SODIUM-REDUCED SEA SALTS

The disadvantage of potassium chloride is that its bitter taste can be observed in the finished meat product and the majority of people can detect this bitter taste if around 3 g of potassium is present per 1 kg of meat product. Three grams of potassium equal around 6 g of potassium chloride given that potassium chloride consists of 52% potassium.

If sodium chloride has to be replaced with potassium chloride, around 15% more potassium chloride has to be applied to dissolve protein to the same extent: this is due to the different proportions of sodium or potassium bound to chloride. Chloride ions are mainly responsible for activating protein, and potassium chloride demonstrates a nonchloride part of 48% compared with 39% in sodium chloride. As a result, more potassium chloride has to be added to end up with the same concentration of chloride within the meat product. People with heart problems must not consume high levels of potassium chloride as excess levels of potassium can cause heart irregularities.

4.3.1 Sodium-Reduced Sea Salt

Sodium-reduced sea salt with a sodium content of only 1.8% is available on the market, but, again, care has to be taken from a food safety point of view when substituting common salt in fermented dried salami. Those salts contain significant levels of materials such magnesium and potassium chloride as part of their natural composition. Usually those types of salt contribute a metallic bitter taste to meat products when applied in excess of 4–5 g per 1 kg of meat product. If sodium reduction is salami is a target, minor steps should be taken in regards to replacing sodium chloride with materials such as sodium-reduced sea salt, and trials should be conducted so as not to sacrifice food safety of the end product.

4.4 CASEINATE AND WHEY PROTEIN

Caseinate is produced from defatted milk and is frequently applied to high-quality salami at a level of 5.0–15.0 g per 1 kg for its contribution to the flavor in the end product. The flavor originating from proteins within caseinate matches well with flavors obtained via proteolysis from meat proteins and result in a typical mellow salami flavor. Proteins within caseinate complement flavors obtained as a result of proteolysis from meat protein, creating a unique mellow and pleasant flavor. The milk is heated to around 45°C, and casein is separated by the addition of acid (citric or carbonic acid). The casein is then washed, an alkaline (sodium hydroxide) is added, and the mixture is heated to around 85°C, after which the slurry is dried. The resulting material is soluble caseinate (sodium, potassium, or calcium caseinate).

Caseinate accounts for around 80% of the total protein found is milk and demonstrates excellent fat-emulsification properties. The protein content of caseinate is around 90%, and it is fully denatured at around 110–115°C.

4.4.1 Whey Protein

Similar to caseinate whey protein adds to the flavor of dried salami products and is added at 5.0–20.0 g per 1 kg of salami. Whey protein is the water phase of milk once casein has been removed. Its flavor is generally equivalent to that of meat, and it matches the light color of poultry meat well. Most commonly, whey protein concentrates of around 35% protein are used in meat products.

4.5 SOY PROTEIN

Soy proteins are generally not applied in salami except in applications where a gel is produced out of soy protein isolate and water (see Section 7.1.2.1 under Chapter 7: Fermented Salami: Non–Heat Treated). Soy protein is by far the most commonly applied protein in the meat industry. Soy is not a new food ingredient, having been used for thousands of years in food. The early Chinese, for example, used soy to make tofu, a product that is still produced today. The word *tofu* means "meat with no bone," and the level of protein in tofu is around 9%. A soy bean contains around 18% oil, 39% protein, 15% insoluble fiber (dietary fiber), 16% soluble carbohydrate (sucrose), around 15% moisture, as well as a tiny amount of ash and other compounds. Soy protein has excellent water-binding properties and is a reasonably effective emulsifier of fat. It also has an extremely high biological value and is easy to digest, and there are some health benefits of consuming soy protein.

In the past, soy proteins had a bad reputation as the addition of soy to meat products had a negative impact in many cases on the flavor as well as the color of a meat product. Those shortcomings, though, are things of the past, and the highly sophisticated soy proteins on the market today are of light color and have no, or very little, impact on the taste and color of the finished product. The beany taste was the result of high levels of raffinose and stachyose in the bean, and due to biotechnological changes, the amount of those substances in soybean today are greatly reduced. Soy isolates exhibiting a protein content greater than 90% are used for the production of a soy gel.

Production of soy products.

1. Selected and cleaned soybeans are cracked to remove the hull and turned into full-fat soybean flakes.
2. The oil is removed with a solvent, and the flakes are dried to obtain defatted soy flakes. Defatted soy flakes can be ground into soy flour, from which textured-vegetable-protein (TVP) can be produced. Defatted soy flour has a protein content of around 52% (dry basis).
3. To manufacture TVP, the defatted soy four is mixed with warm water and a slurry is obtained. This slurry is then minced or pressed through differently sized blades to obtain granules of different sizes, which are subsequently dried. The slurry is also commonly colored to obtain meat-colored TVP granules.
4. Soy concentrate is obtained by removing the soluble carbohydrates from the defatted soy flakes. Soy concentrate has a protein content of around 70–72% (dry basis) and consists basically of protein and insoluble (dietary) fiber.

5. By removing both the insoluble fiber as well as the soluble carbohydrate from defatted soy flakes, soy isolate is obtained which has a protein content of 90–93% (dry basis). Globulins such as glicinin and beta-glicinin are the major functional components of soy protein, as concerns the emulsification of fat and gel formation, respectively. The level of protein in a soy product correlates with the ability of the product to emulsify fat and bind water.

The process is summarized as follows: soybean⇒oil removed⇒defatted soy flour⇒soluble carbohydrates removed⇒soy concentrate⇒insoluble fiber removed⇒soy isolate.

4.5.1 Gelation of Soy

Gelation is the capability to form a gel by producing a structural network, and substances such as water, carbohydrates, lipids, and others can be immobilized within this three-dimensional network. A gel made out of soy isolates is a three-dimensional matrix, and water is held within this matrix. Soy concentrates, though, do not form a gel as the presence of the insoluble fiber inhibits the gel formation: soy concentrates only form a paste. There are major rheological and functional differences between a gel and a paste.

Soy protein was, and still is, one of the food substances at the center of the GMO (genetically modified organism) debate, and leading producers of soy protein have strict quality control measurements in place to guarantee that they are producing non-GMO products. Whereas many countries are comfortable with the use of GM soy, others, such as Europe and Australia, show great resistance toward GM soy protein. Discussions about GM food are often very emotional and unfortunately not often based on facts. Contributing to this situation is that even today there are no reliable and proved studies completed that clearly indicate the effects of GM food on human health. The authenticity of non-GM soy products is commonly tested using polymerase chain reaction (PCR) tests and/or the entire production of the soy proteins may be covered by an identity-preserved (IP) program. In a PCR test, a small string of DNA is replicated many times, and modification, as well as the level of modification, of the protein can be seen. Most soy producers also have an IP program in place in which all stages of the manufacturing process (such as seeding of the bean, harvesting, transport to the processing plant, and processing of the soybean) are closely checked and monitored to ensure that non-GM soy never comes in touch with GM material and cross-contamination is excluded.

The threshold level of GM material within non-GM material is primarily determined by the level of detectability, dependent on the latest analyzing technology available. Nowadays, the presence of GM material can be detected reliably and with statistical back-up to a level as low as 0.9%. Based on those technical capabilities, most countries have adopted this level. Therefore, in these countries, if a soy isolate contains less than 0.9% of GM material, it is called non-GM but not GMO-free as the term GMO-free would require zero tolerance and the presence of absolutely no GM material, which cannot be reliably verified.

4.6 PORK RIND POWDER

Pork rind powder is occasionally added to dried salami products to increase the firmness of the product at a given Aw value as well as to reduce the water activity directly. Typical dosage levels vary between 3.0 and 8.0 g per 1 kg of salami and the water-holding-capacity (WHC) of pork protein is between 1:5 and 1:10. Pork rind powder is an animal protein originating from pork rind (skin) and is a completely natural material. The process to obtain pork rind powder includes thermal and mechanical treatment of the skin. The particle size from pork rind powder ranges from 3 mm down to a super-fine powder and contain around 10–13% fat as well as around 84% protein. If stored at ambient temperatures for a prolonged period of time, the relatively high level of fat in the powder may lead to rancidity. Pork rind powder is very labeling-friendly, meaning that in many countries it does not need to be mentioned on the label of the end product as the material originates from the carcass of a pig in the first place.

4.7 SUGARS

Sugars are carbohydrates, consisting of carbon (C), hydrogen (H) and oxygen (O) and are used in meat products because of their contribution to flavor and their role in browning during the frying process. In raw fermented salamis, sugars are introduced as food for starter-cultures and are predominantly fermented into lactic acid. It is this lactic acid that primarily causes the drop in pH value in salamis during fermentation.

Sugars can be divided into mono-, di-, and oligosaccharides. These compounds are collectively known as sugars due to their sweet taste: a polysaccharide, such as starch, is not sweet and is therefore not a sugar. The muscular sugar glycogen, despite being referred to as a sugar, is actually a polysaccharide. Monosaccharides are the most simple sugars, containing three to seven carbon atoms in each molecule, and are the only form of sugar that can be fermented by starter cultures directly into lactic acid. Pentoses such as ribose, xylose, and ribulose are monosaccharides with five atoms of carbon in each molecule. Hexoses, such as fructose, galactose, glucose, mannose, and tagatose, are monosaccharides containing six carbon atoms in each molecule (Fig. 4.3).

Almost all monosaccharides and their derivates with a ketone or aldehyde group within their molecule have reducing properties (there are a few exceptions such as derivates of aldonic acid). An aldose (polyhyrdoxyaldehyde) is a monosaccharide with a carbonyl group (C=O) on the end carbon forming an aldehyde group (—CHO). If a carbonyl group is on an inner atom, predominantly on the C_2-carbon, forming a ketone, a ketose (polyhydroxyketone) is the result (Fig. 4.4).

Most monosaccharides can be present in chain or ring form but are present primarily in ring form due to intermolecular changes resulting in a ring configuration. A five-sided ring such as ribose is known as a furanose, while a six-sided ring such as glucose is called pyranose. Monosaccharides are either present in their L or D form, depending on the placement of the OH group by the asymmetrical C-atom farthest away from the carbonyl group. The reference substance for the L or D form by monosaccharides is D-glycerinaldehyde. If the OH group is on the right hand side (as it is in glycerinaldehyde), the D form is present ("D" originates from the Latin word *dexter* meaning "right hand side"). If, on the other hand, the OH group is located on the left hand side, the L form is given (Fig. 4.5).

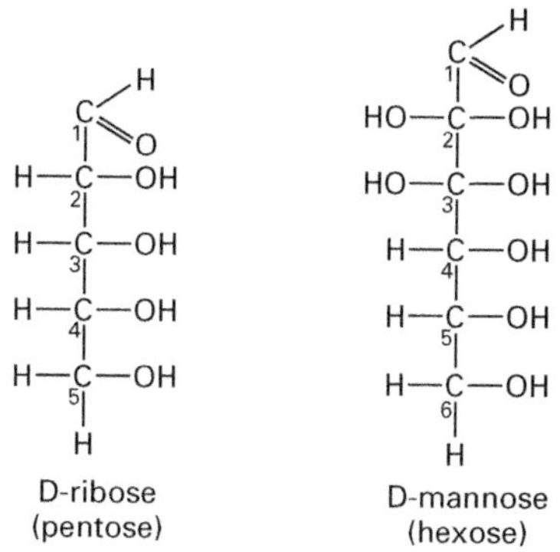

FIGURE 4.3 D-Ribose (pentose) and D-mannose (hexose).

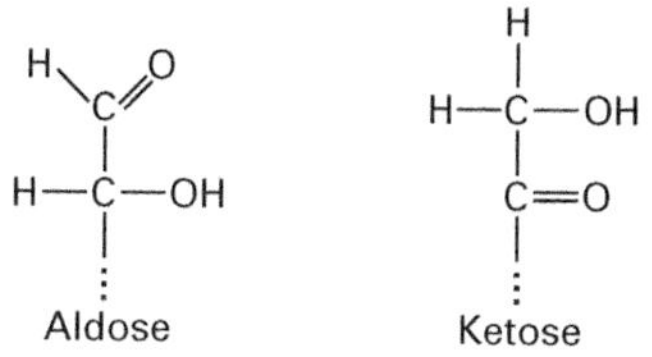

FIGURE 4.4 An aldose and a ketose.

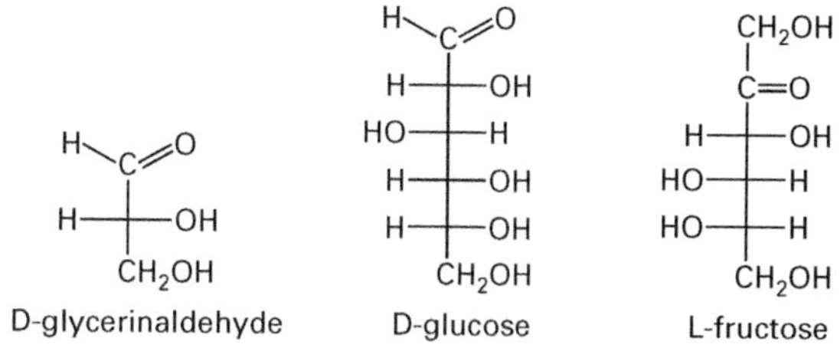

FIGURE 4.5 D-Glycerinaldehyde as well as the D-glucose (aldose) and L-fructose (ketose).

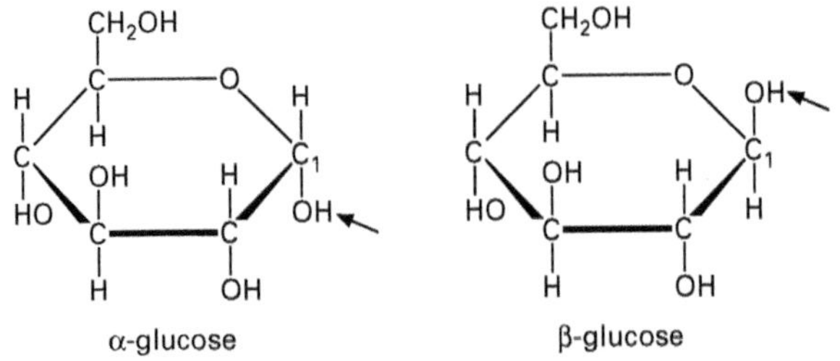

FIGURE 4.6 α- and β-Glucose.

Many sugars only differ in the orientation of the hydroxyl groups (—OH) within their molecules, but this small change causes large differences in taste and melting point as well as in biochemical properties.

Therefore, in sugars such as D-glucose, a distinction is made between α- and β-glucose, depending on the position of the hydroxyl group (OH group) on carbon 1 in relation to the primary alcohol group —CH$_2$OH connected to carbon 5. In α-D-glucose, the OH group on carbon 1 is on the opposite site to the —CH$_2$OH group on carbon 5, while the OH group in β-D-glucose is on the same side as the —CH$_2$OH within the ring-shaped molecule (Fig. 4.6).

The most common form of sugars used in salamis are:

- Glucose (dextrose) is a monosaccharide (hexose) and is the building block for starch as well as glycogen and is also part of sucrose. Glucose is often used in meat products to round-up the flavor as well as for the reason that glucose shows a much greater osmotic pressure within a solution thus reducing the Aw more as, for example, sucrose. Also, glucose can be directly fermented by most types of starter cultures within the production of raw fermented salami, and it is by far the most often applied form of sugar within the production of fermented salami.
- Lactose is a disaccharide consisting of D-glucose and D-galactose units. Cow's milk contains around 4–5% of lactose and human breast milk around 5.5–7.5% lactose. Lactose is, if so, most often used in salami mainly due to its compatibility with meat flavor contributing to a well-rounded and mild flavor in the end product.
- Sucrose (saccharose) is a nonreducing sugar and is made of one unit of D-glucose and one unit of D-fructose. It is used infrequently in salami as it cannot be directly fermented into lactic acid (Fig. 4.7).

4.8 ANTIMOLD MATERIALS

The two most common materials applied to avoid growth of unwanted mold are natamycin and potassium sorbate. Natamycin (pimaricin) is produced by *Streptomyces natalensis*, whilet potassium sorbate is the salt of sorbic acid.

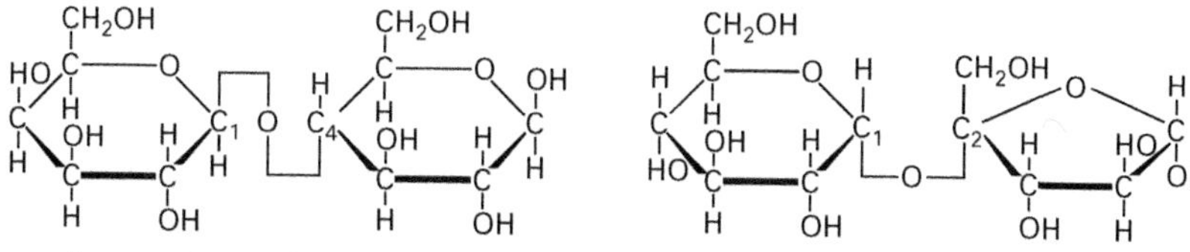

FIGURE 4.7 Lactose and sucrose.

Sorbic acid is 2,4-hexadien acid and acts against mold only in its undissociated form. In meat and meat products, pH values around 5–6 are generally found and around 40% of these acids are present in an undissociated state at such pH levels acting as a preservative. Sorbate is regularly utilized in a dipping solution during salami manufacture. The freshly filled salami is dipped for a few seconds into a 5–10% solution of potassium sorbate, and this dipping process prevents to a large degree growth of unwanted mold during fermentation. Another option is to soak the casings to be filled in a 5–8% solution of potassium sorbate before filling. Natamycin shows poor solubility in water but only 8–10 ppm is needed to avoid growth of unwanted mold.

4.9 MONOSODIUM GLUTAMATE

Flavor enhancers are substances that have no taste or flavor of their own but stimulate the surface of the tongue and enhance the formation of saliva. As a result, the flavor of food in presence of a flavor enhancer is perceived in a more pronounced way. It is believed that the Chinese used the flavor enhancer monosodium glutamate (MSG) as early as 2000 BC, and it has since been used in food products for centuries. MSG is the monosodium salt of glutaminic acid, one of the amino acids, and consists of water, sodium and glutamate itself. Glutaminic acid also functions as a building block for almost all proteins, plays an important role in building muscular structure, and is an interim product in numerous metabolic processes. It occurs naturally in plants such as seaweed, soybeans, and sugar beets, and MSG is usually made by fermenting corn sugar, starch, or molasses from cane or sugar beet.

To obtain the desired effect as a flavor enhancer, glutamate must be present in its L-configuration. In food with a pH value of 4.5–8.0 (all meat products fall into this range), if MSG is present in a concentration of 0.2–1.0%, a pleasant, slightly sweetish as well as salty taste will develop, enhancing the food's own taste. The flavor-enhancing effect of MSG is ultimately based on the level of dissociated glutaminic acid within a product and the level of dissociation depends on the pH value. Glutaminic acid is completely dissociated by a pH value of 7.1 and is only dissociated to around 40% at a pH value of 4.6. In meat products, where generally a pH value of 4.6 (low pH for a fermented salami) up to around 6.2 is found, the flavor-enhancing effect of MSG becomes more pronounced at a higher pH value.

In addition to the four traditional taste perceptions (salty, sweet, sour, and bitter), the evasive "umami" taste was created to best describe the contribution of MSG to the taste of food. This specific "umami" taste plays a vital role in the acceptance of food in Asia, with umami being interpreted as savory or delicious. MSG enhances meaty and salty flavors, reduces bitter notes, and has no or very little impact on notes of sweet and sour in meat products or other foods. MSG is believed to be responsible for the "Chinese restaurant syndrome" with symptoms of a slight headache, nausea, chest pain, and tingling in the facial areas as well as the neck. Only around 1.5% of the total population, though, is susceptible to this kind of syndromes and then only if quite a high amount of MSG is consumed on an empty stomach. At typical usage levels of around 0.5–2 g/kg of meat product, the consumption of MSG, even on an empty stomach, hardly causes any problems.

4.10 RIBONUCLEOTIDE AND OTHER FLAVOR ENHANCERS

As a replacement for MSG, flavor -enhancers such as inosine-5-monophosphates of sodium, potassium, and calcium (E 631–E 633) are used. These flavor enhancers are applied at a very low level (0.007–0.05%) in the final product but are frequently mixed with MSG to achieve the specific umami taste (MSG taste), and applied on their own, no umami taste is produced. Examples of flavor enhancers in this group are disodium guanylate (E 627) and disodium inosinate (631), also known as 5′-guanylate monophosphate (GMP) and 5′-inosinate monophosphate (IMP).

GMP and IMP together are known as ribonucleotide and carry the E-number 635. Ribonucleotide have a 40–50 times stronger flavor enhancing activity compared with MSG and are naturally present in meat at around 0.02%. A mixture of 19 kg of MSG with 1 kg of ribonucleotide provides the same flavor-enhancing effect as 100 kg of MSG on its own as a strong synergistic effect is seen between those two materials.

Taurin, a semiessential amino acid, was once promoted as a flavor enhancer but was never commercially successful. Alapyridaine, a flavor enhancer isolated from beef broth, enhances the taste of sweet, salty, and other piquant tastes, influencing more than one sense of taste at the same time. It has not yet been applied in meat products, though.

4.11 WATER/ICE

Water as such is not an additive but it fulfills major technological functions when producing cooked salami. A molecule of water is made of two hydrogen atoms and one oxygen atom. By weight, the oxygen is 8 times heavier than the hydrogen within a water molecule and the two atoms of hydrogen are bound to the oxygen at an angle of 108 degree. Because the hydrogen atoms are bound to oxygen at an angle, water has its highest density at +4°C and therefore ice floats on water. Water is a polar molecule and in polar molecules, there is an unequal distribution of charges. The polar nature of water makes it a suitable medium for living cells.

Different levels of hardness are found in water and hardness can be divided into temporary or permanent hardness. The degree of hardness of water is determined by recalculating the amount of calcium (Ca^{2+}) or magnesium (Mg^{2+}) present in water into "milligram of calcium oxide per liter of water" (mg CaO/L). One degree of German-hardness (1°dH) equals 10 mg of CaO in 1 L of water, which is equivalent to 7.15 g of Ca^{2+} or 4.34 g of Mg^{2+} per liter of water. Other countries may use other figures and other scales to express hardness of water but the basis is the amount of Ca in mg/liter of water (Table 4.1).

Temporary hardness is caused by the presence of calcium or magnesium hydrogencarbonate (HCO_3^-), which evaporates out of water on heating. Permanent hardness is found when the calcium or magnesium is not present as hydrogen carbonate but as chloride, nitrate, phosphates, or sulfate, and these substances remain in the water after heating.

From a technological point of view, to dissolve protein, the presence of water is required to act as a solvent in conjunction with phosphates and salt on muscular protein. During the manufacture of cooked salami, cold water (or ice) is used to maintain a low temperature during processes such as cutting and mixing given that the activation of muscular protein is most effective at low temperatures. Low temperatures during various processing steps also reduce the risk of bacteria growth.

The water used in meat products must be of drinking quality (potable) and the amount of chlorine within water should be low. Enhanced levels of chlorine react with nitrite present in cured meat products and a significant amount of nitrite could be lost, which would not be available for the formation of curing color anymore.

TABLE 4.1 Total Hardness (Permanent and Temporary) of Water Expressed in mg CaO/L

Degree of Water Hardness	Total Hardness (mg of CaO per L of H_2O)
Very soft	0–40
Soft	40–80
Slightly hard	80–120
Medium hard	120–180
Hard	180–280
Very hard	Above 280

4.12 SPICES AND SPICE EXTRACTS

Spices are plants or parts of plants, which are added to food for their contribution toward flavor, aroma, and taste. They are not added for nutritional purposes as they do not supply energy and are consumed in very small quantities. Around 50% of the spices produced in the world are used in the meat-processing industry, and all spices used in meat products are of premium quality. Second-grade quality spices are generally bought by consumers in the supermarket. Besides the four basic tastes (bitter, salty, sour, and sweet), flavor of food is the sensory impression of food obtained in the mouth. On the other hand, aroma is experienced by the mucous membrane within the nose and over 5000 different aromas can be distinguished.

Spices have an impact on the flavor and appearance of a product as well as its taste. Some spices also aid digestion and increase appetite overall; some, such as rosemary, sage and their extracts, have antioxidative properties; while others, such as thyme and garlic, have bacteriostatic properties. The bacteriostatic properties of garlic and thyme, though, cannot be exploited to the full in meat products, as the inclusion level required for a satisfactory bacteriostatic effect would not be tolerated by the consumer from an organoleptic point of view. In total, around 60 spices demonstrate bacteriostatic properties to some degree.

Most spices used are dried, frozen, or preserved in brine. Spices should be stored under cold or ambient temperatures and the storage area should be dry. The relative humidity (RH) in the storage area should not exceed 65%; if stored too warm, aroma components in the spices vaporize and are lost. Stored under high humidity, mold can grow, and if stored unprotected against the impact of light, color can fade, and atypical aromas can develop. Once a bag of spice is opened, it is vital to close the bag properly afterward as volatile oils and other aroma compounds will evaporate. Storage in moisture and air-proof containers or aluminum bags is an advantage. Spices can be highly contaminated with pathogens and spores, and when purchasing spices, the specification should show figures regarding the total-plate-count (TPC), the *Salmonella*, *E. coli*, and *Listeria* count, and so forth.

Spices originate from roots, barks, leaves, herbs, blossoms, seeds, or fruits of certain plants. The different types of flavors originate commonly from various alcohols, aldehydes, esters, lactones, and ketones. Spices also contain water, salts, and low-molecular substances such as sugar, fatty acids, as well as amino acids. They also contain high-molecular substances such as lignin, cellulose, and starch. Depending on the type of spice, they also contain volatile oils, bitter, spicy, and color components. The level of essential oil in spices varies dramatically and is between 0.01% and 16%; the average is around 1%. The level of extracts within spices varies greatly and is between 0.1% and 24%, with most spices containing extracts at an average between 2% and 5%. Most spice blends in meat products are applied at a level of 3–6 g/kg of product based on a spice blend containing 100% spices. However, spice blends are frequently a mix consisting of natural spices as well as some extracts. Spice blends based on extracts only are added at a very low level per 1 kg of meat product. Commonly, spice blends contain filler -materials such as sugars and salt, and strict guidelines are in place regarding the terms used to describe spice blends containing different levels of nonspice components. For example, in most countries, a product sold as a spice blend can contain nonspice materials such as carriers in the form of sugar or salt up to a maximum level of 5% without having to declare the filler materials.

Different parts of plants are the origin for spices and herbs:

1. Seeds like mustard, coriander, cardamom, and nutmeg
2. Fruits such as pepper, paprika, caraway, and pimento
3. Leaves and herbs like marjoram, rosemary, oregano, mint, sage, and thyme
4. Spices from blossoms, parts of blossoms, or buds such as saffron and clove
5. Roots and rhizomes like ginger, turmeric, and horseradish
6. Bark spices such as cinnamon and cassia
7. Bulb spices such as onion and garlic.

Hot flavors come from chili, mustard, and pepper, while pungent flavors are found in spices such as anise, cardamom, and cloves. Sweet flavors are present in allspice, cinnamon, and vanilla, while tangy flavors are seen in ginger and sumac. Amalgamating flavors are present in fennel and coriander seed. The spicy-hot component in spices such as chili, belonging to the capsicum annum spices, is capsaicin, while the nonspicy pepper flavor is based on piperine. For chilies, the rule of thumb is the smaller the chili, the more heat is in it. Capsaicin is present primarily in the seed of a chili as well as inside the chili wall and is around 300 times more spicy-hot compared with piperine. Chili contains up to 0.7% of capsaicin, and the level of heat in spices is generally expressed in Scoville heat units (SHUs).

Spices such as onions and garlic only develop their characteristic and pronounced flavor once cell walls are destroyed through cutting or chewing given that an enzyme, present in those spices or fruits, has to come in touch with other intrinsic substances to produce the flavor. If a small onion would be swallowed as a whole, no onion flavor would be created as the enzyme did not get a chance to react with other substances to form the actual onion flavor. The same applies to garlic. The combination of the enzyme allinase in onions with derivatives from amino acids (present within an onion) and the presence of syn-propanethial-oxide, which converts sulfoxide (amino acids) within the onion into sulfenic acid, come in contact with air, thus irritating the lachrymal glands and causing the crying sensation (lachrymatory factor) in the eyes. Due to their pyruvic acid content, onions can cause browning in meat products, and it is the pyruvic acid that is also responsible for the pungency of onions. The acid acts also an oxidizing agent, and enzyme activity also plays a role in discoloration in meat products based on the presence of onion. The problem can occur from fresh, dried, or even frozen onions, but if onions are heat treated upfront, the problem is reduced. As onion and garlic, chives belong to the allium family (allium = onion plants). Onion, or *Allium cepa*, exhibits only around 180 g of essential oil per 1000 kg of fresh material, while garlic produces around 2.5 kg of essential oil per 1000 kg of fresh material. Garlic, or *Allium sativum*, contains up to 25% of sugar, while onion contains around 10% of sugar. Different levels of sugar play a vital role within the browning of those materials on heat treatment during the Maillard reaction. The predominant aroma component in garlic is allicin.

Pepper is used in the forms of white, black, green, and pink pepper. Black peppercorns are the green and unripe peppercorns picked from the vine and dried in the sun. Upon drying in the sun, the outer husk turns from green into black and the volatile oil piperine is formed during this process, which gives black pepper its characteristic flavor. White pepper is obtained by soaking the almost ripe peppercorn in water and subsequent removal of the outer husk before finally being dried. White pepper is "hotter" compared with black pepper but the piperine taste in white pepper is significantly less pronounced as in black pepper. Green peppercorns are picked green and placed either in a salt brine or freeze dried to prevent an enzymatic reaction, which would turn the green into a black corn. Pink pepper is the ripe berries put into a salt brine to avoid any enzymatic reaction. Pink pepper has a fruity note and underlying heat.

Real vanilla is very expensive and therefore vanillin is commonly used instead. Vanilla is obtained by drying and curing the tasteless and green pod from a tropical orchid native to South America. Vanilla contains around 1.5–3% of actual vanillin and 0.15–0.3% hydroxy-benzaldehyde as its characterizing flavor component. This ratio between vanillin and hydroxy-benzaldehyde must be 10:1, and based on this ratio, the addition of synthetic vanillin to vanilla can be detected. Vanillin can also be made chemically and is not a spice but tastes very similar to vanilla.

It is carotenoids that give paprika its red and yellow color. Carotenoids are soluble in fat and if used excessively, they can give the fat portion in a sausage a yellow appearance. Excess use of oleoresins from paprika also results in a touch of yellow color in sausages. Spices and herbs can also be a source of nitrite as well as nitrate, which can cause pinking of meat products where no pinking is desired, but this does not present a problem in salami as salami contains nitrite anyway.

The level of microbiological contamination in untreated spices varies dramatically, which can present a microbiological risk in the production of fermented dried salami products. Country of origin, harvesting methods applied, hygiene during transport, and other parameters determine the level of contamination, and the bacteria count in black pepper can be as high as 10^8 cfu/g. Treatment of spices with different materials such as ethylene oxide is common to reduce the bacteria count, but an increasing number of countries do not permit the use of ethylene oxide anymore. Steam treatment is frequently applied as an alternative to reduce the bacteria count but is by far not as effective as treatment with ethylene oxide. Most leaves and herbs cannot be treated with steam at all, though, given that appearance and taste changes dramatically. Generally, steam-treated spices and herbs should demonstrate a total plate count (TPC) of less than $10^5/g$, less than $10^3-3 \times 10^3$ on yeasts and molds, and less than 50 of *E. coli* per gram. *Salmonella* should be not detected in 25 g. The micro-count of spices and herbs is critical toward the intended application of spices in meat products. Meat products that do not undergo heat treatment during manufacture, such as raw fermented salami, fresh sausage, or marinated meats, should be produced by using either treated spices, untreated spices presenting a low bacteria count, or spice extracts. Such microbiologically sensitive products should not be produced using untreated spices showing a high initial bacteria count given that a faulty product and a greatly reduced shelf-life would be the result. The selection of such low-bacteria-count spices is not that critical for meat products that are fully cooked during manufacture. Other options for a reduction in micro-count would be the treatment of spices with chlorine, UV light, irradiation, formaldehyde, hydrogen peroxide, or high pressure. Most of those treatments did not make a successful transformation into the commercial world or are not permitted.

Different spices demonstrate different stability against heat. Heat stabile spices are sage and chilies, whereas less heat stabile spices are rosemary, thyme, cardamom, paprika, clove, and pepper, and other spices such as coriander, pimento, mace, marjoram, and ginger are not heat stabile. The lower stability toward heat is compensated through a higher addition rate to meat products during processing.

Spice oleoresins or extracts consist of essential oils, soluble resins, and other materials present in the original spice as well as nonvolatile fatty acids. An oleoresin or extract is basically the "total flavor of a spice in a concentrated form," obtained via extraction in the first place and treatment with solvents afterward. Solvents commonly applied are alcohols as well as ether and extracts do not contain cellulose or starch, which makes those materials almost sterile.

Essential oils are the principal or specific flavoring component of spices mostly obtained via steam distillation out of the original spice. The vast majority of a spice, whether it is a fruit, the bark, a seed, or any other part, does not contribute much to the flavor given that those parts are predominantly made of fibrous tissue and water. Only paprika, sesame seed, and cayenne pepper have a flavor on their own, and the content in essential oil is not of significant importance toward their flavor. Liquid-soluble spice flavorings are commonly blends of oleoresins as well as essential oils and diluted to a specific strength by the addition of a solvent such as glycerol. Oleoresins and essential oils demonstrate the advantage of exhibiting a standardized and uniform flavor, are low in bacteria count, and demonstrate a long shelf-life. Flavor development within meat products by applying oleoresins takes place faster compared with the use of natural spices and full flavor strength is obtained right away. When using natural spices, it takes around 12–24 h until the full flavor strength in a meat product is developed, but the flavor lasts longer within the product if natural spices are used compared with oleoresins. Therefore, the evaluation of the flavor and taste of a meat product such as cooked salami, produced with natural spices, should take place once the product is at least 1 day old.

4.13 HYDROLYZED VEGETABLE PROTEIN

Hydrolyzed vegetable proteins (HVPs) are applied worldwide for flavoring meat products at a level of 0.2–4.0 g per 1 kg of meat product. The main sources for HVP are soy, maize, or gluten, and soy is the preferred source for meat-flavored HVPs. During the manufacture of HVPs, plants such as soy get broken down into their individual components such as amino acids and other peptides by the help of acids, mostly hydrochloric acid (HCl). HCl is afterward neutralized with an alkaline such as sodium hydroxide (NaOH), and salt, water, and sodium salts of amino acids are obtained. The resulting slurry is called HVP and is subsequently exposed to heat treatment, which causes different amino acids to react with each other. New and different flavors are the result of reactions under the impact of heat and HVP is comparably inexpensive.

When plants are broken down into their individual components (amino acids) by the help of acids such as HCl, though, the chloride ions from those acids react with lipids found in those plants (the lipids are found in trace amounts only). A reaction between chloride and lipids creates 3-monochloropropane-1,2-diol (3-MCPD), and this substance has been shown to be carcinogenic in laboratory animal studies. Once formed, the stability of 3-MCPD depends on the pH value and temperature to which substances are exposed. High temperatures and increased pH values in meat products accelerate the degradation of 3-MCPD. As a consequence, it is a common trend currently in certain countries that manufacturers of meat products are using more often flavors and flavorings based on yeast extracts, which are free of MCPD.

4.14 ANTIOXIDANTS

Antioxidants must work in low concentration such as 200–500 ppm per 1 kg of meat product, should demonstrate good solubility in fat or fatty material, must be nontoxic, and must not change or alter the taste of the meat product itself. Antioxidants in meat products have the primary task of deactivating or neutralizing free radicals to slow down the development of rancidity. More specifically, antioxidants extend the period of time until a significant number of oxidation-related substances are formed and rancidity is observed. Deactivating peroxides is a secondary function of antioxidants and of far less importance. Antioxidants can block radicals, which are substances with a lone pair of electrons and therefore "negatively" charged (see Section 1.9 under Chapter 1: Meat and Fat), by donating hydrogen (H*) and stabilize such radicals. Most antioxidants become radical themselves by donating H* but are significantly less reactive and stabile compared with radicals obtained from autoxidation. As a result of deactivating radicals, less hydrogen peroxides are formed, which can fall apart into substances such as aldehydes as well as ketones and therefore contribute to the rancid flavor and taste. Substances such as ascorbic acid are donors of H*, which block a radical, and the newly formed antioxidant radical itself is stable. Ascorbic acid also acts as an oxygen scavenger at the same time and therefore deactivates oxygen otherwise used, first, for the formation of peroxide and, second, breaking bonds on double linkages in unsaturated fatty acids, which leads to the formation of radicals.

Phenolic substances such as tocopherol, carnosic and rosmarinic acid, and phenols from smoke (all cyclic unsaturated systems) can also deactivate radials as well as hydroperoxides by either donating H* (phenols are weak acids) or neutralizing them by absorbing radicals into their ring-shaped molecule. When a phenol donates H* to neutralize hydroperoxide, the phenol is oxidized to a phenol radical, which can be reduced again by gaining hydrogen. The donor of the hydrogen can be ascorbic acid and the phenol radical is reduced again to phenol. Ascorbic acid can donate two H* and is therefore a very active donor to phenol radicals. Within such a system, ascorbic acid acts as an indirect antioxidant. On the other hand, phenols can absorb radicals into their ring-shaped molecule and the configuration of the ring does not change within the process. The absorbed radical is bound, or neutralized, within the ring by moving the radical around inside the ring but not being bound to a specific atom.

Ascorbate, the salt of ascorbic acid, is insoluble in fat and shows very little antioxidizing effect in fat. Once introduced into meat, which has a slightly sour pH value by nature, some undissociated ascorbic or erythorbic acid is obtained, and those compounds can also donate H* to block radicals. More effective antioxidants on fat would be tocopherol, butylated hydroxianisole (BHA), or butylated hydroxitoluene (BHT), which are all phenolic substances.

Tocopherols are a group of four of eight naturally occurring compounds, which are fat-soluble vitamins known collectively as vitamin E. The position of the methyl group within their ring structure makes the difference within those four materials, which are known as alpha (α)-, beta (β) -, gamma (γ)-, or delta (δ)-tocopherol. In meat and meat products, α-tocopherol is predominantly used given that it is significantly less expensive compared with γ- and δ-tocopherol. However, antioxidative properties shown by α-tocopherol in meat products are very limited, while γ- and δ-tocopherol exhibit much stronger antioxidative properties. Commonly, a mix of γ- and δ-tocopherol in a ratio of 1:1 is applied, which shows a syngergistic effect as an antioxidant. As a result, antioxidative properties from tocopherols are contrary to their vitamin character given that α-tocopherol demonstrates the strongest vitamin character but is the weakest antioxidant within the group of tocopherols. Tocopherols deactivate free radicals in fats given that tocopherols are fat soluble and therefore delay rancidity in meat systems containing fat. At the same time, tocopherols are insoluble in water and cannot be used in ham brines. The usage rate of tocopherol is around 0.02–0.05% or 0.2–0.5 g per 1 kg of finished product. Excessive levels of tocopherol turn the antioxidative function upside down and the antioxidant acts pro-oxidative and will speed up rancidity.

Polyphenols in green tea oil such as catechine are stronger antioxidants than, for example, vitamins A, C, and E but very expensive and leave a green tint in meat products. Rosemary and oregano contain phenolic substances such as carnosic acid. Carnosic acid donates H^+ to neutralize a free radical and turns into carnosol, which is itself an antioxidant by deactivating free radicals. Rosmanol, another antioxidant, is formed out of carnosol in the next step and then another antioxidant galdosol is obtained out of rosmanol. This chain reaction explains that carnosic acid acts as an antioxidant in many different ways and substances such as rosmanol and carnosol demonstrate only around 40% antioxidative properties compared with carnosic acid. Rosemary and oregano extracts are already very effective at usage levels of 0.05–0.08 g/kg of finished product, and oregano extract has an even stronger antioxidative impact compared with rosemary extract. Those phenolic compounds also have an impact on other gram-positive bacteria as well as on *Listeria monocytogenes*.

Sage demonstrates an antioxidative impact as well, and both rosemary and sage can delay the period of time until fat-containing meat products exhibit rancidity around 14-fold. Rosemary and sage contain around 0.3% of carnosic acid, which surpasses tocopherol as well as BHT in their functionality. Carnosic acid is in most countries not permitted as a direct food additive but is present in rosemary, sage, and oregano extract. Those phenolic substances, as said, either absorb radicals into their ring-shaped molecule or donate H^+ to stabilize free radicals. Synthetically made phenolic antioxidants such as BHT and BHA are very similar to each other and not permitted in some countries. BHA is a mixture of two isomers where one of them, 2-BHA, is present at around 82–85%, and the other one, 3-BHA, at around 15–18%. BHT or BHA are very rarely applied in meat products and if so, use is made in products such as sausages and meatballs; the usage rate is commonly 0.01% based on the fat content as a single antioxidant or 0.02% if both are applied in combination. Today, BHT and BHA are under suspicion of causing cancer.

Heavy metal ions contribute to and speed up oxidation. Citric acid can form a complex with heavy metal ions (chelating agent) and heavy metal ions lose their supporting potential to oxidize fat. In animal fat itself, heavy metal ions are rarely present, but in a finely comminuted sausage, such as a hot dog, heavy metal ions coming from meat are finally present in the sausage mass and the presence of citric acid is welcome.

4.15 NATURAL SMOKE

Smoke has been applied for around 80,000 years in the production of meat products and is produced by incomplete combustion (pyrolysis) of wood material such as sawdust or wood-chips. Friction and steam condensation are two other methods of generating smoke. In the friction method, a piece of wood of a certain size is pressed against a fast running rotor with a ripped surface and high friction forces are the result. Friction smoke is obtained, which exhibits a high level of phenols, carbonyls, and other acids while showing a low level of tar and polycyclic aromatic hydrocarbons (PAH). Pyrolysis takes place through the heat obtained during friction between the piece of wood and the fast rotating rotor with the ripped surface. Steam smoke is obtained by the use of overheated low-pressure steam, and the temperature within this process is between 300 and 450°C. A spiral-shaped element transports sawdust into an area where such overheated steam is applied to sawdust or wood-chips, causing pryolysis. The generated smoke is cooled down to around 85°C until it reaches the smoking chamber, which causes an increase in moisture (RH) within the smoke and this gives the basis for the name "steam-smoke." Steam-smoke is basically free of tar as well as PAH.

Wood consists of around 50% cellulose, 25% hemicellulose, and 25% lignin. From those materials, around 50–70% turn via pyrolysis into smoke, and pryolysis results in burnable coal rather than ash by burning wood slowly at a certain temperature. The term "hemicellulose" describes polysaccharides that are made of different pentoses and hexoses.

Pryolysis takes place in four steps:

1. Drying of the wood up to around 160°C
2. Pyrolysis of the hemicellulose between 180 and 250°C
3. Pyrolysis of the cellulose between 250 and 300°C
4. Pyrolysis of the lignin between 300 and 550°C

The optimal temperature for combustion is between 350 and 500°C, and temperatures below or above this range causes a considerable higher amount of unwanted substances within smoke afterward. The most well-known and dangerous one at the same time of those unwanted substances is 3,4-benzopyrene, which is carcinogenic and belongs to the group of PAH. If smoke is generated at temperatures between 350 and 500°C, PAH contamination is greatly reduced and concentrations of less than 1 ppb (parts per billion) per 1 kg of smoked meat product are obtained.

Smoke is a highly complex mixture of gas-like substances, solid particles (particulate phase), as well as water, and around 600 components within smoke are known today. The particulate phase accounts for around 80%, while the gaseous phase for around 20% and the gas fraction is not visible to the human eye yet is very complex. The composition of smoke depends primarily on the type and moisture content of wood used as well as on the method used to generate smoke. Main components of smoke having the greatest impact on meat products, are phenols, organic acids, and carbonyls. Most of those substances are in the gaseous phase and not in the particulate phase.

Phenols are not, as one might think, alcohols but rather acids (carbolic acid) and are obtained through the pryolysis of lignin at a temperature between 300 and 450°C. Phenols can release a proton from a hydroxyl group and that makes them a weak acid. The visible particle fraction consists of small, liquid colloidal smoke particles. Those particles are very small in size, around 1 μm, and are distributed within the gas fraction. Much larger particles such as ash and tar are also part of the visible particle fraction. Two phenolic substances coming from lignin, syringol or guaiacol (2-methoxy phenol), are of importance regarding the smoke color and flavor in meat products. Guaiacol results from lignin present in soft wood and is not favored as it gives a dark-dull smoke color as well as a rough and unpleasant smoke taste. The desired golden-brown smoke color, as well as pleasant smoke flavor, comes from syringol, which is obtained from lignin in hardwood such as oak and hickory.

Wood-chips or sawdust should be stored in a dry area and no animals should have access to this area. It happens often that wood-chips are contaminated with animal feces and/or urine. Sawdust or wood-chips have to be moistened first with water; otherwise, the wood material gets too hot very quickly (above 450°C) and little smoke is obtained. The amount of water added to the wood material is around 20–30% from the weight of the wood material. Also, the smoking chambers must not be overloaded during the application of smoke to avoid hot-spots. Air still has to move freely within the smoking chamber to achieve an even drying, smoking, and cooking impact (Table 4.2).

The main functions of smoke are:

1. Development of the smoke color

Smoking of meat and meat products results in an appealing golden brown color, which is very attractive to the human eye. Carbonyls are the main color-forming agents, and carbonyl is absorbed into the slightly moist surface of the product. Subsequently, carbonyls react with amine to form the desired smoke color. To a small degree, phenols also contribute to smoke color. It is important to have the correct and an even level of moisture on the surface of the salami to obtain an appealing and even smoking color. If the surface is too wet, or unevenly wet, a brownish color and even streaking ("tiger stripes") can be the result of smoke absorbed unevenly on the surface of the product.

TABLE 4.2 Three Methods of Smoking

Method	Temperature (°C)	RH (%)
Cold smoke	15–25	50–85
Warm smoke	25–50	50–80
Hot smoke	50–90	30–85

Streaking occurs when there is some free moisture on the surface of the product, which is a risk when cold smoke is applied to salami in the early stages of fermentation.

2. Development of the smoke flavor

It is well known that smoked meat products taste differently than nonsmoked products. Substances such as formaldehydes, lactones, and up to 20 different phenols (guaiacol and syringol) are primarily responsible for the smoke flavor. Hardwood such as maple, oak, beech, hickory, and mahogany is preferred given that those types of wood give a clean and nontarry flavor.

3. Smoke as preservative

Smoked meats and meat products last longer, which has been well known for thousands of years. Formaldehyde, phenols, and acetic acid are the main agents for the impact on extending the shelf-life of smoked products given that these materials are very effective antimicrobial substances.

Phenols are acids that denature proteins and disrupt cell membranes. Disrupted cell membranes eventually kill the cell or make it very hard for the cell to survive or even to grow.

4. Smoke adding to bite

Smoked sausages such as frankfurters get a much better bite, or snap, through smoking; however, this fact is of little importance in salami. Components of smoke, mainly formaldehyde and other organic acids, combine with the activated protein on the surface of the noncooked sausage and subsequent thermal treatment creates a firm layer around the sausage, which is largely responsible for the snap. Smoke contains around 0.6–1.0% formaldehyde.

The most common mistakes during smoking of salami are:

1. An uneven smoke color is obtained due to faulty equipment and insufficient smoke distribution within the smoking chamber. Insufficient or uneven drying of the product to be smoked is commonly a reason for uneven smoke color given that the remaining amount of moisture on the surface interferes with the absorption and incorporation of smoke into the surface. The product can be spotty, exhibiting tiger stripes, and the color will be dull overall.
2. Overloading of the smoking camber leads to uneven air flow and can cause uneven smoke color.
3. A negative impact on the flavor can be seen by using moldy sawdust or wood-chips. Sawdust or wood-chips also must be free of wood-impregnation substances.
4. The first application of smoke should only take place when development of curing color within the product is completed given that acids present in smoke interfere with the development of the desired curing color.

4.16 LIQUID SMOKE

Liquid smoke is produced by burning selected woods under controlled conditions. The smoke obtained is condensed in water and recycled until the desired concentration is given. Liquid smoke can be applied to meat products via atomization, a dipping or shower system, brine addition to meat products, and in the form of smoke-impregnated casings. Atomization is the spraying of liquid smoke under a predetermined pressure through a nozzle into the smoking chamber and creating a cloud of smoke made of very tiny particles.

Before liquid smoke is introduced into the smoking chamber onto salami, the surface of the product has to be dry for proper absorption of the smoke applied. After atomization, the cloud of smoke is allowed to dwell for around 10–15 minutes within the chamber before a drying step of 5–10 minutes is introduced. The spraying process can be repeated several times until the desired smoke color is obtained, and two or three applications of spraying results in a nice golden-brown color.

Meat products that should exhibit smoke flavor without being smoked experience the addition of smoke, in the form of a powdered smoke flavor or liquid smoke, directly into the product during the manufacturing process. Care has to be taken that smoke flavor, containing acids from smoke, does not come into direct contact with nitrite causing chemical reactions between acids (from smoke flavor) and nitrite resulting in nitrose gas and loss of nitrite. Casings are also available for cooked salami that demonstrate a layer of smoke at the inside. Such casings are generally not soaked before filling as such treatment with water during soaking would wash out the layer of smoke. The product is filled into the casing, and a 1-h period of drying by low humidity and temperatures around 70–75°C is commonly the first step to fixate the color on the meat product before being finally cooked by the application of water or steam.

In the past, liquid smoke commonly produced a slightly bitter taste on meat products, but due to ongoing improvements on the quality of liquid smoke, this negative side effect belongs in the past.

Liquid smoke shows several advantages compared with natural smoke:

1. Because liquid smoke is "standardized," an even smoke color on the finished products can be obtained all the time.
2. There is no emission of smoke into the air and therefore smoking with liquid smoke is environmentally friendly.
3. Smoke chambers are easy to clean given that liquid smoke does not contain tar and other tacky substances. Most of the time, the smoking chambers can be cleaned with water only and no chemical cleaning detergent is required.
4. Liquid smoke is practically free of PAH.

4.17 COLORS

A large number of colors are permitted to be applied in food overall, but only a few are suitable for meat products and colors are barely applied when producing salami. Colors that restore or enhance the red curing color are used, and the acceptance of colors in meat products varies dramatically from country to country.

Within the European Union and other countries, the maximum level of color permitted in a meat product is not specifically defined but expressed with the Latin term "quantum satis," meaning "as much as needed."

Colors can be divided into three major groups:

1. Natural colors from plants or animals such as carotene, cochineal, beet red, annatto, and curcumin (orange-yellow color derived from the rhizome of tumeric)
2. Colors derived from natural sources such as caramel and titanium dioxide, a white pigment
3. Artificial colors such as tartrazine, sunset yellow, or brilliant blue

The most commonly applied colors in meat products are as follows.

1. Carmine de Cochineal (E 120)

Carmine is the red color that accumulates in the shell of pregnant scale insects (*Dactilopius coccus*). A liquid extract is obtained from such dried female insects and then mixed with alumina to produce the alumina solution of carminic acid, which is the main coloring agent in carmine. Since the alumina solution of carminic acid is not water soluble, it must first undergo treatment with material such as ammonia or carbonate (alkalines) to obtain water-solubility. Cochineal dissolves well in water and is applied widely in cooked hams, cooked sausages, as well as salami. As cochineal is not soluble in fat, the color is well suited for salami and ham with fat and skin-on as it provides color to meat only and does not discolor fat at the same time. A coloring effect in a meat product can be observed when adding as little as 0.02 g of cochineal per 1 kg of meat product by the coloring being stable against light, pH variations, and thermal treatment.

2. Beet red (betanin)—(E 162)

It is the extract of beet root and is purple-red. The main coloring agent is β-D-glucopyranoside of betanidine (betanin), but beet red shows poor stability against the impact of light and heat.

3. Fermented rice

Fermented rice (angkak) is of rice origin and a dark-red powder. It is in most countries not classified as a color per se but frequently applied for coloring purposes and is made from a natural pigment present in mushrooms such as *Monascus purpureus* and *Monascus angkak*, which grow on moist rice. Such azaphilone pigments are ultimately extracted from *Monascus*. At this stage, fermented rice in not permitted within the European Union and some other countries given that the mycotoxin citritin is present in fermented rice and citrinin is known to cause cancer. Most types of fermented rice start to show an impact on the color of a meat product when applied at 0.1 g per 1 kg of product. Fermented rice is stable against light and heat.

4. Paprika oleoresin

Paprika oleoresin is widely used in salami-type products such as pizza salami or in any other type of "spicy-hot" salami as the orange-red color is synonymous for "spicy-hot" end products. The oleoresin is not a color per se but the main reason for being introduced is the color-giving effect in sausages. Several types, or qualities, of paprika oleoresins are available, and the concentrations vary from 20,000 to 100,000 color units (CU). Generally, the better the quality of the oleoresin, the longer the color lasts in the meat products. An appealing and genuine paprika-red color can be obtained in products by adding around 0.1–0.3 g of a 40.000 CU oleoresin per 1 kg of product.

4.18 CHEMICAL NONBACTERIAL ACIDULANTS

Besides acidification being the result of starter cultures fermenting sugars into acids that reduce the pH value in salami, chemical acidulants are applied for reducing the pH value in salami without having to rely on bacteria activity.

4.18.1 Glucono-δ-lactone

Glucono-δ-lactone (GDL) is a derivate of glucose (ie, a saccharic acid) and is a ring-shaped molecule. GDL exhibits six carbon (C) atoms, and an OH group is attached on every C atom (Fig. 4.8).

In meat technology, GDL is predominantly used as acidity regulator in fermented salami and occasionally as a color enhancer. Basically, 1 g of GDL lowers the pH value within salami by 0.07–0.09 pH unit. GDL itself is a whitish powder and is not sour. Once in contact with water, coming from meat, the ring-shaped molecule opens up and hydrolyzes to gluconic acid. This acid causes the decline in pH value within fermented salami, and GDL is commonly applied at levels between 3 and 12 g/kg of salami.

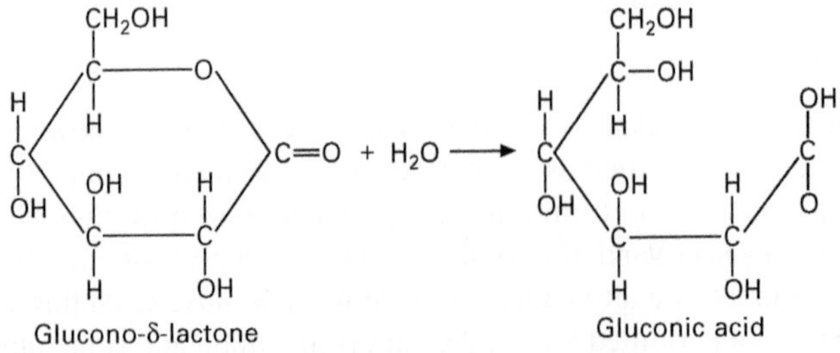

FIGURE 4.8 GDL and gluconic acid.

4.18.2 Encapsulated Citric Acid

Citric acid anhydrous carries the E number 330 and the chemical formula is $C_6H_8O_7$. This food acid is occasionally used in the production of salami for acidification purposes. Generally, 1 g of citric acid/kg of salami lowers the pH value by around 0.2–0.3 pH unit and acidifies a salami therefore around 3–4 times stronger compared with GDL. Citric acid also acts significantly faster compared with GDL in regard to acidifying a salami given that the acid is added directly into the product and that no conversion from salt to acid, as is the case with GDL, is required. Citric acid also acts as a chelating agent by binding heavy metal ions such as copper as well as iron and therefore acts as a secondary antioxidant, but the contribution to being an antioxidant is marginal. Citric acid anhydrous, despite the fact of not being highly hygroscopic, tends to lumping (caking) if stored above an RH of 75%. The monohydrate material tends even more to lumping because of the higher internal level of moisture. Both materials are chemically stable if stored at room temperature at around 20°C and are nontoxic. Citric acid is a naturally occurring fruit acid and commercially produced by the fermentation of a carbohydrate material. It is a white, odorless crystal material with a strong acid taste and is freely soluble in water and ethanol. Coated, or encapsulated, citric acid is used mostly in semicooked or fully cooked salami, which also should exhibit an "acidified" note, and acidification takes place primarily during the heat treatment of the product. Most materials used for encapsulation melt around 45–50°C, so time is provided first for proper development of curing color before heat treatment takes place adding the desired acidified note to the end product. Encapsulated citric acid is most often applied in fast-acidified salami products such as pizza salami where the entire production process is very short and cost driven.

4.19 FIBERS

Fibers are a recent addition to salami to accelerate the speed of drying as well as reducing the Aw within a salami product, and powdered cellulose was the first type of fiber applied because it is one of the most abundant organic materials on earth. Through the addition of fibers into a salami a secondary "pipeline" system is created, allowing faster removal of water without obtaining case-hardening. As a result, drying time of a salami can be reduced as much as 20–25% creating an interesting economical factor given that drying is a highly time-consuming process. Fibers are usually added at around 5–15 g per 1 kg of salami, and various sources of fibers are applied. Wheat fiber is made out of the straw of wheat and is white and has very little taste. Because it is made out of the straw (cellulose), it is claimed to be gluten free.

Fibers also reduce the Aw in salami due to their WHC, which is, on one hand, based on the "capillary effect." The capillary effect can be explained in a way that if a straw is put into water, the level of water inside the straw is higher as the water level around the straw—it is a kind of a "sucking effect." The WHC of a fiber is generally tested using a centrifuge-system were oversaturated (overhydrated) fibers are placed in a centrifuge and the amount of water held within the fiber is divide by the amount of fiber used in the first place. WHC of fibers is also influenced by the level of pectin present in fibers since pectin is highly hydrophilic. Citrus fiber exhibits high WHC due to high levels of pectin present. WHC is also influenced by the length of the fiber particles, and longer particles generally show higher WHC. The diameter of fibers also plays a role toward WHC, and fibers with a smaller diameter usually show increased WHC. Besides citrus fiber, fibers from carrot and potato also show similar high WHC. WHC from soy and pea fiber is slightly less compared with citrus, carrot, and potato.

Chapter 5

Color in Cured Meat Products and Fresh Meat

Given that the consumer buys with his or her eyes, an attractive and stable color in meat and meat products has a major influence on the buying decision of the consumer. Much research has been done and continues to stabilize the color of fresh meat and to investigate the use of nitrite in cured meat products.

5.1 RETENTION OF COLOR IN FRESH MEAT AND UNCURED MEAT PRODUCTS

The color pigment of the muscle tissue is myoglobin, while hemoglobin is the color pigment of the blood. The level of hemoglobin in meat depends largely on the degree of bleeding during the slaughtering process. Generally, the color in raw meat is made from about 90–95% myoglobin and about 2–5% hemoglobin given that lean meat is never totally free of blood. Hemoglobin exhibits a quaternary structure with four polypeptide chains (globin), each containing a heme group. The molecular weight of hemoglobin is 65,000 Da, and the molecular weight of myoglobin is 17,000 Da. Myoglobin is therefore around 4 times smaller than hemoglobin. Some other proteins contribute to the color of meat as well but on a very small scale and can be neglected.

Myoglobin is a monomeric globular protein and is made from a protein, the colorless globin, with a color-giving heme group. The heme group consists of a flat porphyrin ring exhibiting a central iron atom (Fe^{2+}). This iron atom has six coordination links, called ligands, and four of those six ligands are attached to atoms of nitrogen while one is attached to the globin. Substances such as oxygen, water, or nitric oxide (NO) may bind to this sixth ligand, and the state of oxidation of the sixth ligand plays a vital role in the color of fresh meat.

Myoglobin is present in fresh meat in three different forms, which are in equilibrium with each other. Those three forms are reduced myoglobin, oxymyoglobin, and metmyoglobin, and the "final" color of fresh meat is always a mix of these three forms of myoglobin. As long as the central iron core is present in its reduced state as Fe^{2+}, reduced myoglobin or oxymyoglobin is present. Reduced myoglobin is purple-red in color and water is bound to the sixth ligand. Meat packed under vacuum and the center of pieces of meat exhibit such reduced myoglobin. Oxymyoglobin is bright-red and oxygen bound to the sixth ligand. Low temperatures support the formation of oxymyoglobin given that, first, solubility of oxygen is enhanced at low temperatures and, second, little oxidizing activity from enzymes can be seen as well. Metmyoglobin is obtained if the central iron atom is oxidized to Fe^{3+} by a loss of an electron (oxidation) and water is present on the sixth ligand. Metmyoglobin is brownish-gray in color and is mostly present in areas of low O_2 concentration between the oxygenated outer layers of meat and anaerobic inner areas of meat. Such color in meat is commonly seen in meat displays, which is not attractive to the human eye.

Metmyoglobin cannot absorb oxygen directly, and enzymes present in meat have to reduce metmyoglobin first to reduced myoglobin. Subsequently, reduced myoglobin can again take up oxygen to form the red oxymyoglobin. If meat is stored for a prolonged period of time, this enzyme activity comes to an end and metmyoglobin obtained can no longer be reduced to reduced myoglobin. As a result, no more oxymyoglobin can be formed, and the color of meat remains brownish-gray. Even though a very thin layer of red oxymyoglobin is present on the surface of raw meat stored for a prolonged period of time due to the impact on oxygen on those surface layers, the majority underneath is metmyoglobin and the brownish-gray color of metmyoglobin sooner or later dominates the meat color overall.

The overall color of raw meat is greatly determined by the amount of metmyoglobin present. Meat demonstrating up to 30% metmyoglobin is still of intensive red color, and even at a concentration of 30–45% metmyoglobin, the color of meat is still red. Having 45–60% of metmyoglobin causes a brownish-red color, and by 60–75% of metmyoglobin, the color of meat is reddish-brown. By concentrations of above 75% metmyoglobin, the color is brownish-gray. As seen, the state of the globin plays an integral part in the color of raw meat. As long as the globin is present in its native form and the iron ion is present in its reduced Fe^{2+} state, regardless of whether water or oxygen is bound to the sixth ligand, the color of meat is purple or bright-red (see Table 5.1).

TABLE 5.1 Different States of Myoglobin

	Oxidation	Sixth Ligand	Color	Globin
Myoglobin (reduced)	Fe^{2+}	Water	Purple-red	Native
Oxymyoglobin	Fe^{2+}	Oxygen	Bright red	Native
Metmyoglobin	Fe^{3+}	Water	Brown	Native
Denatured globin	Fe^{3+}	Water	Gray	Denatured

The color of fresh meat also depends on the light-scattering ability of meat itself. PSE meat exhibits a small myofibrillar volume and demonstrates high light-scattering ability. As a result, such meat appears pale. On the other hand, DFD meat shows a large myofibrillar volume and low light-scattering ability, which allows light to penetrate deep into the muscle. Therefore, DFD meat is dark-red. The concentration of myoglobin in grams per kilogtam of lean meat follows the order: beef⇒lamb⇒pork⇒poultry, with poultry exhibiting the least amount of myoglobin. Hence, different cuts of the same animal demonstrate different levels of myoglobin as well.

Generally, muscle tissue from male animals contains higher levels of myoglobin than does meat from female animals. Also, muscles heavily used for movement such as shoulder and leg require more oxygen, which is transported primarily via red blood cells, resulting in a darker color of muscle tissue compared with muscles such as breast (in poultry). Hence, myoglobin concentration of heavily used muscles is generally higher than in muscles used less for movement. Animals fed in a stable that therefore hardly move around generally exhibit a lighter meat color compared with free-range animals, which move around all day.

This also explains, for example, why chicken thigh meat is darker than chicken breast meat or why pork shoulder meat is darker than pork loin. The terms "dark" and "light" meat are commonly used to describe these differences in color. Also, increased age of an animal causes an increase in the concentration of myoglobin, thus making meat darker.

Packed nonoxygenated fresh meat must be covered by a highly permeable packaging-film in order to secure a sufficient supply of oxygen for reduced myoglobin to form oxymyoglobin again, and oxygen permeability of packaging films should be around 6–7 L per m^2/day. Materials such as low-density polythenes are used for this type of packing films.

Vacuum-packed meat is of an unattractive purple color because the presence of a vacuum (nonpresence of oxygen) results in reduced myoglobin. The red and desired oxymyoglobin is restored quickly once the packaging is opened and the meat is exposed to oxygen again.

Oxygenated modified-atmosphere packed (MAP) fresh meat, though, should be covered with material exhibiting low permeability to maintain a high level of oxygen within the packaging. Hence, gas such as CO_2 is introduced in combination with oxygen to suppress bacteria growth.

Generally, MAP fresh meat contains around 70–80% oxygen and around 20–30% of CO_2. Those high concentrations of oxygen, besides being needed to maintain oxymyoglobin, favor the growth of aerobic bacteria such as *Bronchothrix* and *Pseudomonas*, which shorten the shelf-life as both are aerobic bacteria and oxygen present is used as food. The introduction of CO_2, on the other hand, keeps bacteria growth under control. CO_2 forms carbonic acid in conjunction with water ($CO_2 + H_2O \Rightarrow H_2CO_3$) coming from the meat itself and disrupts cell membranes.

Carbon monoxide (CO) must be kept away from fresh meat in case a non-pink color is wanted given that CO binds to myoglobin to form carboxy-myoglobin, which is red. The affinity of CO to bind to hemoglobin is around 300 times greater than the affinity of oxygen binding to hemoglobin, thus explaining the toxicity of CO by inhibiting oxygen to bind to hemoglobin. CO exhibits a high affinity to myoglobin as well as hemoglobin and creates a strong cherry-red color in meat, which can last for up to 3 weeks and is therefore occasionally used in MAP fresh meat because it creates a strong and lasting red color. However, due to its toxicity, CO is not permitted in most countries. In countries where CO is permitted, gas mixtures of 60–70% of CO_2, 20–30% of nitrogen, and 0.3–0.4% of CO are applied for fresh meat. The addition of small amounts of CO enhances the color as well as having a positive impact toward shelf-life. High concentrations of CO, though, would cause a green color in meat. Another risk in the use of CO is that meat, spoiled from a microbiological point, still has a nice color and therefore the consumer might be mislead into thinking that the product was fresh.

In a typical gas mixture containing CO, the high levels of CO_2 provide good shelf-life and nitrogen is used as a filler gas to replace oxygen. By using a CO gas mixture, an attractive color is produced by the impact of CO and the shelf-life is good due to the presence of CO_2.

Methods are practiced to bypass the problem that CO is, in many countries, not permitted by using "tasteless smoke." Legal authorities in most countries see this as a way of adding CO to meat in an indirect, and illegal, way. Smoke, generated at high temperatures, is subsequently filtered to remove the particular phase of smoke as well as all smoke flavor–giving components, resulting in a material containing, besides some nitrogen complexes and oxygen, CO_2 as well as CO.

The introduction of CO-containing tasteless smoke to packed fresh meat creates a lasting red color without adding CO directly as an additive. Even though the amount of CO introduced this way is of no harm to human health, the consumer might be mislead by the term "tasteless smoke," which actually stabilizes the red color and has no connection to smoking of meat. The effect of such tasteless smoke on the color of fresh meat is comparable to around 0.4% of CO present in MAP meat.

In summary, in MAP foods, the gases normally applied such as CO_2 and nitrogen fulfill the task of replacing oxygen and therefore remove the oxygen required for growth of aerobic spoilage bacteria as well as delay rancidity. CO_2 is the active component against microbiological spoilage because it forms carbonic acid in conjunction with water from the meat. Nitrogen fills the gap to exclude oxygen.

In spoiled meat, the color pigment may be decomposed and globin separates from the heme group. As a result, the phorphyrin ring is destroyed and the iron core separates from the heme. Ultimately, such meat will green as a result of the presence of green-colored choleglobin. Aerobic bacteria can cause discoloration as some produce H_2O_2, which is a strong oxidizing agent, thus destroying the color pigment and resulting in a green or pale color. Species belonging to *Pseudomonas* can cause green and bluish colors as metabolic byproducts. Hence, some bacteria produce hydrogen sulfite (H_2S), which leads to the formation of sulfmyoglobin and resulting in green-colored meat as well. Sulfmyoglobin can also be obtained in vacuum-packed meat at high pH values, generally at or above 6.2. Micrococci can cause discoloration, creating an atypical red color. Peroxides present in fatty meat, originating from fat autoxidation, can also transform myoglobin into metmyoglobin by oxidizing the iron core from its Fe^{2+} into Fe^{3+} configuration, resulting in greening of the meat.

5.2 NITRITE AND NITRATE

Cured meat products such as salami should demonstrate a strong and stable red curing color given that the customer buys "with his/her eyes." To obtain a stable red curing color in meat products, sodium nitrite ($NaNO_2$) is generally the material of choice. $NaNO_2$ dissociates into Na^+ as well as NO_2^- and nitrite is the water-soluble, highly toxic salt of nitrous acid (HNO_2). For nitrite, and here specifically sodium nitrite, the lethal dose for humans is about 1.1 g. For this reason, a mixture of salt and nitrite is commonly sold to manufacturers of cured meat products. The level of nitrite within such salt–nitrite mixtures varies generally between 0.5% and 20%, depending on the country's regulation.

Nitrite is a strong oxidizing agent and can cause pinking in noncured meat products at levels as low as 3–5 ppm of nitrite/kg of meat product. Occasionally, potassium nitrite is used in applications such as low-sodium products. Based on the difference in molecular weight, around 30% more potassium nitrite has to be applied for the same impact on color development to be achieved compared with sodium nitrite.

Nitrate (NO_3^-) is the salt of nitric acid (HNO_3) and is also readily soluble in water. In meat products, potassium nitrate (KNO_3) is the most commonly applied source of nitrate, although in a large percentage of all cured meat products nitrite is the material of choice. Potassium nitrate dissociates into K^+ and NO_3^- ions and is introduced sometimes in products such as long-dried fermented salami and cured air-dried products. Nitrate does not contribute directly to the formation of the red curing color, and nitrite is the active component regarding the formation of the wanted red curing color. If nitrate is applied, it has to be reduced first to nitrite, which then forms the curing color in conjunction with myoglobin, the color of meat.

Nitrite is used in meat products for the following reasons:

1. Formation of the red curing color
2. Formation of the curing flavor
3. A bacteriostatic impact
4. Acts as an antioxidant

In cured meat and meat products, a strong and stable red color is required and nitrite is the only material of choice. To this day, no substitute is found for nitrite when it comes to generating a stable and appealing curing color. To obtain a strong and stable curing color in cured meat products, around 30–50 ppm is required per kilogram of meat product. However, 3–5 ppm of nitrite in a meat product is sufficient to create a pinkish touch, which is generally not wanted in products such as roast pork (roast pork in most countries is a noncured meat product) or other "white" products such as chicken loaf. Therefore, great care has to be taken with products that should not be cured to ensure that no traces of nitrite are introduced.

The curing flavor originates from reactions between NO and the numerous substances naturally present in meat, such as aldehydes, alcohols, inosine, and, of great importance, several sulfuric components. Different carbonyl complexes also result from the presence of nitrite and contribute to the curing flavor. Around 30–50 ppm of nitrite is required for the development of the typical curing flavor.

The antioxidative effect of nitrite is based on the fact that nitrite oxidizes to nitrate as well as forming solid complexes between the iron core of myo- and hemoglobin and therefore reduces the number of free iron ions (Fe^{2+}). Free Fe^{2+} ions and some other pro-oxidative materials are bound by nitrite delaying the development of rancidity. Around 20–60 ppm of nitrite are needed for nitrite to act as an indirect antioxidant. NO binding to myo- and hemoglobin during the formation of curing color does not give oxygen the chance to bind to myo- and hemoglobin and oxidation of myo- and hemoglobin is reduced as well.

Nitrite also acts as a preservative against bacteria such as *Salmonella*, *Staphylococcus aureus*, and especially *Clostridium botulinum*. A concentration of 80–140 ppm of nitrite/kg of meat product is an effective hurdle against growth of those bacteria, especially in canned and retorted products. However, nitrite has relatively little impact against Micrococci, *Lactobacillus*, and Enterococci.

The impact against potential food-poisoning bacteria is one of the major reasons why nitrite is still permitted as an additive, as a ban of nitrite would most likely increase the number of severe cases of food poisonings. Nitrite does not accumulated in the human body, and most humans do not consume nitrite-containing food on a daily basis. The impact of nitrite as a hurdle against bacteria growth in cured and cooked meat products is generally overrated mainly because 80–140 ppm is hardly present in the cooked product given that most food standards in place do not permit levels as high as 140 ppm in a cooked product. Other hurdles, such as a low initial bacteria count of the meat itself, sufficient cooking, or thermal treatment of the product and storage of the cooked product at low temperatures, exhibit overall a much greater impact on the shelf-life of a meat product as purely the presence of nitrite. The fact that nitrite is present in a cooked product is better than it being absent, but it should not be seen as the only integral part to ensure microbial safety. Obtaining proper levels of nitrite in the cooked product is of greater importance toward the maintenance of a strong curing color than being the sole hurdle against unwanted bacteria.

A hint of pink color in noncured meat products can be the result of nitrite/nitrate present in spices or herbs or water used during the manufacture of products. Such materials used for the manufacture of meat products can contain nitrite, but more often nitrate, and up to 70 ppm can be seen per liter of water. Nitrate is reduced in the uncooked meat product to nitrite, and formation of the red curing color starts to take place. Also, an unwanted touch of pink in noncured meat products can be obtained from combustion of coal or natural gas. Some byproducts of direct-fired gas ovens, especially nitrogen dioxide (NO_2), penetrates the outer layers of a noncured product and causes pinking given that NO_2 is water soluble. Hence, nitrite cross-contamination between nitrite and non-nitrite products within a factory must be avoided to eliminate the risk of obtaining a pink color in an uncured meat product.

Excess levels of nitrite in a meat product lead to discoloration known as nitrite burn, and meat exhibits a green color as a result. Nitrite burn is commonly seen at levels of nitrite above 600 ppm of nitrite/kg of meat, and the formation of nitrihemin, a green-brown pigment, is responsible for this process. Hence, nitrite burn is also partly connected to the pH value in meat products and low pH values reduce the level of nitrite causing nitrite burn. Also, high levels of nitrite result in high levels of HNO_2 during the conversion of NO_2 into nitric oxide, and this temporary acid denatures myoglobin, which supports the formation of a green/yellow color caused by nitrite burn in the first place.

The presence of nitrite in meat products can cause the formation of nitrosamines, which can cause cancer; however, nitrosamines are only obtained if nitrite is present in a meat product at the same time as secondary amines. Secondary amines are very seldom found in meat products, and the amount of nitrosamine formed in cured meat products is therefore relatively low. During the formation of nitrosamine, nitrite forms nitroso groups ($-N=O$), which react with secondary amines under the impact of heat. In cured meat products such as cooked sausages and cooked hams, the level of secondary amines is next to nothing and therefore nitrosamines cannot be obtained. Also, to obtain nitrosamine, the pH value of food must be below 5.6, and of all meat products, only fermented salamis exhibit such pH levels in the finished product. Finally, temperatures above 140°C (grilling) are required to form nitrosamines. Based on that, products such as pan-fried bacon are a potential but small risk only if heavily fried under high temperatures until basically all fat is melted and a fat- and water-free material is obtained. A fat- and water-free material such as that would support the formation of nitrosamines if it were exposed to temperatures above 140°C for a prolonged period of time. Raw fermented and acidified pizza salami (pH value around 4.6–4.8) is also a risk product but, like bacon, only if heavily exposed to high temperatures and if a fat- and water-free product were obtained, which is highly unlikely, as in the case of bacon.

5.2.1 Natural Nitrite

Dried vegetable extracts from various sources such as celery and beetroot contain "natural nitrite." Those materials are applied as a substitute for "chemical nitrite" in products that should exhibit a more "natural" or "less-processed" image. In general, 15,000–25,000 ppm of nitrites is present in 1 kg of such vegetable extracts, which equals 15–25 ppm if applied at 1 g/kilogram of meat product. However, 15–15 ppm of nitrite/kilogram of salami is not sufficient for developing a strong and stable curing color as well as being a hurdle against bacteria during the initial stages of fermentation given that 100–150 ppm of nitrite is generally applied per kilogram of salami. If such levels of ingoing nitrite are achieved through the addition of natural nitrite, the flavor of the salami would be negatively affected. Also, natural nitrite is many times more expensive compared with "chemical" nitrite, but this can be compensated if the consumer is will to pay for the "healthier" image of the end product. The other challenge is to purchase natural nitrite from a source that can guarantee a consistent level of nitrite within the vegetable extract, because "natural" ingredients are generally hard to standardize.

5.3 MECHANISM OF COLOR DEVELOPMENT IN CURED MEAT PRODUCTS

In most applications today, sodium nitrite is used and the application of nitrate occurs rarely given that nitrite is the color-giving component and acts as a hurdle against bacteria growth. Potassium is rarely used in combination with nitrite as potassium nitrite but is frequently applied in combination with nitrate as potassium nitrate. When nitrate is added to a meat product, it has to be reduced first to nitrite to contribute to the formation of the curing color as well as acting against bacteria growth. The reduction of nitrate to nitrite ($NO_3 \Rightarrow NO_2$) is a process accomplished by enzymes as well as bacteria and parameters such as temperature and pH value play a vital role within this reduction process. Nitrate is reduced to nitrite by the enzyme nitrate reductase, and bacteria such as Micrococci produce this enzyme. Nitrate reductase only shows activity if the pH value is above 5.5 and the temperature is above 8°C. Such elevated temperatures speed up the process but cannot be fully applied within meat products as a microbiological risk would be the result. This explains why NO_3 is generally not reduced to NO_2 in a cooked cured meat product once stored at temperatures of 0–4°C and is therefore not available for stabilizing the curing color in a cooked product. Hence, in a fully cooked meat product, the enzyme nitrate reductase is not present (active) because it was denatured during thermal treatment, and residual nitrate in a meat product is basically of no use regarding the stabilization of the curing color.

Once nitrite is obtained from nitrate or when nitrite is added directly to a meat product (as is most commonly the case), the reduction of nitrite to NO

($NO_2 \Rightarrow NO$) depends on factors such as pH value, time, temperature, and the presence, or absence, of color enhancers. Longer periods of time and elevated temperatures favor the formation of NO (as such factors generally speed up chemical reactions) but can be applied only in a limited way within meat products due to microbiological risks. The presence of a color enhancer speeds up the formation of NO as well. NO is an odd electron–numbered, very reactive gas and a strong oxidizing agent. The reduction of nitrite to NO in a sour environment (as in a meat product) is a chemical process and no enzymes are involved. It is a spontaneous chemical reaction based on HNO_2 obtained as an intermediate product. HNO_2 is the conjugate temporary acid of nitrite and is unstable in solution. With a pH above 6.5, HNO_2 is fully dissociated and is present as $H^+ + NO_2^-$ ions. At such high pH levels, no NO can be built from dissociated molecules given that only undissociated molecules of HNO_2 can dissociate into NO, HNO_3, and water. The reaction of HNO_2 dissociating into NO follows the process:

$$3NO_2^- + 3H^+ \Rightarrow 3HNO_2$$
$$3HNO_2 \Rightarrow HNO_3 (H^+ + NO_3^-) + H_2O + 2NO$$
$$NO + myoglobin \Rightarrow nitrosomyoglobin$$

This reaction takes place at pH values below 6.5, and in meat products, a pH value of around 4.7 (salami) to 6.0 is present in the final product and therefore the parameters for the formation of NO under slight acid conditions are given. A pH value of 5.3 would be the optimum given that all HNO_2 is present in its undissociated form at this pH value and the optimum amount of NO would be obtained.

Unfortunately, a pH value of 5.2 is the IEP of the actomyosin complex at the same time. This represents a conflict regarding WHC where high pH values around 6.0–6.4 are desired. As a consequence, formation of curing color and WHC are contrary effects and the full potential cannot be obtained from either in a meat product such as a cooked ham but is of much lesser importance in salami as removal of water is the aim during drying (see Fig. 5.1).

The actual formation of curing color takes place when nitrite, a strong oxidizing agent, oxidizes myoglobin first quickly into metmyoglobin, which is brown-gray. At the same time, NO is obtained from nitrite as described earlier

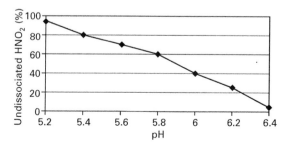

FIGURE 5.1 Shows the level of undissociated HNO_2 in relation to the pH value.

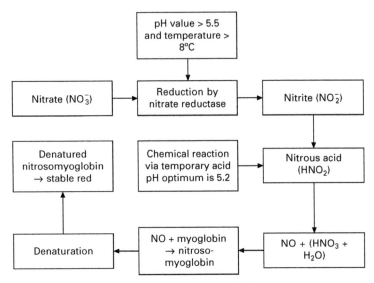

FIGURE 5.2 Mechanism of curing.

via the intermediate step of HNO_2 and binds to metmyoglobin on the sixth ligand, resulting in nitroso-metmyoglobin. Nitrosometmyoglobin is red and exhibits an iron atom, which is charged positively three times (Fe^{3+}). Nitrosometmyoglobin is subsequently reduced to nitrosomyoglobin, which is not fully stable yet. Only after denaturation does nitrosomyoglobin decompose into globin and nitroso-myochromogen, showing the pink and finally stable curing color in meat products (see Fig. 5.2).

In raw fermented acidified salami, nitrosomyoglobin is denatured by the impact of acidification at pH 5.2 (IEP) and below. In slow-fermented and non-acidified salami, as well as in cured and dried meat products, the combination of low Aw with a high concentration of salt denatures nitrosomyoglobin, too.

NO can be obtained in two ways. First, in a slightly sour media such as a meat product, NO is obtained when HNO_2 if formed from nitrite within a chemical process. Second, ascorbic acid directly reduces residual nitrite via HNO_2 to NO following the reaction:

$$2HNO_2 + \text{ascorbic acid } (C_6H_8O_6)$$
$$\Rightarrow 2NO, 2H_2O, \text{ and } C_6H_6O_6 \text{ (dehydroascorbic acid)}$$

The main reasons for poor color development and color stability in meat products are:

1. Insufficient myoglobin in meat, which, unfortunately, cannot be altered
2. Insufficient nitrite introduced into the meat product in first place
3. No, or insufficient, color enhancer (ascorbic acid, ascorbate) used
4. Time provided for color development was too short

The final color of a cured meat product always contains to some degree metmyoglobin as well as some oxymyoglobin, but nitrosomyoglobin is by far the biggest portion. It is never the case that all myoglobin is converted into nitrosomyoglobin and that no met- or oxymyoglobin would be present at the same time. Green dicoloration of cured meat products such as a cooked ham is occasionally caused by the presence of strong oxidants such as H_2O_2 produced by heterofermentative *Lactobacillus* (*Lb. fluorescens*). Very strong oxidation leads ultimately to the formation of a verdoheme complex, which is green. Further oxidation results in a yellowish color. Sufficient heat treatment of cooked salami generally eliminates heterofermentative *Lactobacillus* and core temperatures of 70–72°C are required. Greening and other forms of discoloration in cured cooked meat products are commonly signs of poor hygiene present in the factory. Hence, machines used during the production and slicing of the finished product must be kept clean and free of any cleaning and disinfection materials given that those materials are either very alkaline or acid by nature.

Fading of curing color is also greatly enhanced by the impact of light. Complexes containing NO are susceptible to photodissociation (photolysis), and the impact of light reduces color stability as a result. In cooked cured products, a fair amount of added nitrite is oxidized to nitrate by oxidase enzymes and is no longer available for color development. Poultry products, as well as all other noncured meat products, can exhibit an unwanted pink color by the unintentional contamination of such products with nitrite. Nitrosomyoglobin can be formed as a result of introducing nitrite-contaminated spices, water, and other materials. CO obtained from direct-fired gas ovens can produce CO-hemochrome, and such a substance causes discoloration by forming a nonwanted pink color on the surface on noncured meat products.

5.4 COLOR ENHANCERS

The reaction from NO_2 via HNO_2 finally to NO, as well as binding of the NO to myoglobin, takes place very slowly at the pH levels present in meat (around pH 5.5–6.2). Color formation also takes place at very low temperatures during various processing steps, and low temperatures slow down chemical processes. The addition of substances such as glucono-delta-lactone (GDL), ascorbic acid, ascorbate, isoascorbate, or citric acid speeds up the process of color formation and stabilizes curing color in the finished products. For a faster formation of curing color, all of those materials slightly reduce the pH value in a meat product and higher levels of undissociated HNO_2 are obtained. As a result, more NO is produced and the curing color is enhanced as the result. The function of a color enhancer could also be described as a kick-starter given that nitrite falls apart by itself into HNO_2 by a pH value at ot below 5.7 only. Unfortunately, most meat products exhibit a pH value above that level, thus requiring a substance to start the conversion from nitrite into NO via the formation of temporary HNO_2.

The most commonly applied color enhancers in salami as well as cooked salami are ascorbic acid, ascorbate, and erythorbate (isoascorbate).

Sodium ascorbate is the sodium salt of ascorbic acid, while erythorbate is the salt of erythorbic acid. There is no difference from a technological point of view between erythorbate and ascorbate in a cured meat product regarding the stabilization of curing color except that around 10% more erythorbate has to be introduced compared with ascorbate to show the same impact regarding stabilizing the color. The enhanced addition of erythorbate is based on the difference in molecular weight. Interestingly, ascorbate stabilizes the red curing color in cured noncooked meat products that are packed under vacuum but speeds up discoloration if the same cured noncooked product is stored under the impact of oxygen (not vacuum packed) and light. Under the impact of oxygen, the unstable nitrosomyoglobin turns into a buffer (antioxidant) for NO radicals and ascorbate supports this process. As a result, NO separates from myoglobin and the impact of light speeds up the separation of NO from myoglobin even more.

Ascorbic acid is a strong reducing agent that enables the fast and direct formation of NO from residual nitrite and enhanced levels of nitrosomyoglobin are the result, thus stabilizing the curing color obtained in first place. It is primarily used in cured cooked sausages at around 0.4–0.6 g (400–600 ppm)/kg of meat product but can also be applied in salami at similar levels. Ascorbic acid also reacts with the temporary HNO_2 during the conversion of nitrite into NO, speeding up the formation of NO in this way. Ascorbic acid is vitamin C but in most countries this vitamin image cannot be promoted as a marketing or sales tool. Isoascorbic acid has the same effect in stabilizing the developed curing color but is not a vitamin. Ascorbic acid also acts indirectly as an antioxidant given that it stabilizes hydroxyperoxides, which are obtained by the formation of metmyoglobin.

Greening in cooked salami can be caused by bacteria but can also be due to excess addition of color enhancers such as ascorbic acid. Excess levels of this color enhancer cause the formation of large amounts of HNO_2s within a short period of time, which ultimately causes myoglobin to turn green or, in severe cases, yellow. Excess levels of acids can denature myoglobin prematurely, and extremely pale or slightly yellow/green colors can be seen in the finished product.

Ascorbic acid must never be applied at the same time as nitrite to meat products because nitrite and ascorbic acid react instantly with each other and nitrose gas (nitrogen dioxide) is formed: this can be seen as yellow/brown fumes and is highly toxic. The nitrite required within the meat products is used up if nitrite and ascorbic acid are added together, and discoloration will be seen in the final product due to the lack of nitrite. When producing a cured cooked sausage, nitrite and ascorbic acid have to be added separately into the sausage mass.

Because of its lower molecular weight, slightly more ascorbate or erythorbate has to be applied to every kilogram of meat product to achieve the same effect as if ascorbic acid would be applied. To be more specific, 87 g of ascorbic acid is as effective as 100 g of ascorbate. Ascorbate is generally applied in salami at 0.4–0.5 g/kg of meat product, while erythorbate is generally introduced at

0.5–0.7 g/kg of product. Ascorbate is the salt of ascorbic acid but is significantly more expensive than erythorbate.

Ascorbic acid, or ascorbate, has three major functions in cured meat products. First, it reduces nitrite directly to NO and facilitates the formation of nitrosomyoglobin. It accelerates the formation of the red curing color, which would take place without it as well but at a much slower speed. Second, it stabilizes the curing color by acting as an antioxidant, neutralizing, or deactivating, peroxide radicals in the surface of the product once such surface is exposed to oxygen and UV light. Third, by reducing the level of nitrite in the cooked finished product, it prevents, or reduces, the formation of nitrosating agents such as nitrous oxide (N_2O_3) and thus the formation of nitrosamines.

GDL is occasionally introduced as a color enhancer as well as acidulant at 1–2 g/kg of salami mass as GDL lowers the pH value in the salami mass, and therefore elevated levels of undissociated HNO_2 are obtained. As a result, more NO is formed, which contributes to a stronger curing color.

5.5 MEASURING COLOR: L*-a*-b* SYSTEM

The human eye is a very specialized instrument and can distinguish between 7 million different colors. From those 7 million, only around 3000 have a specific name and only 12 from those 3000 are used in everyday life. The color of meat or of a meat product can be measured and represented using the L*-a*-b*-system.

The L* value represents the difference between white and black: an L* value of 0 is black and an L* value of 100 is white. A positive a* value, or a+ value ranging from 0 to +50, represents the red tone of the product. Higher a+ values indicate a darker-red color. A negative a* value, or a− value, reaching from 0 to −50 represents the green tone of a sample and −50 is the darkest green tone. A positive b* value, or b+ value, ranging from 0 to +50, represents the yellow tone of a sample. A b+ value of +50 is the strongest yellow tone. A negative b* value, or b− value, ranging from 0 to −50 represents the blue tone of a sample and here as well, −50 is the strongest blue tone (see Fig. 5.3).

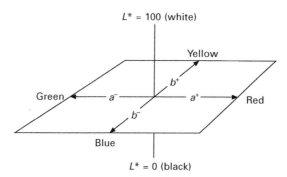

FIGURE 5.3 Three-dimensional structure within the L*-a*-b* system.

Chapter 6

Casings

Because salami has to lose weight during fermentation, drying materials are chosen that are permeable to air and moisture. However, some low-cost salami-type products are put into waterproof casings, which do not allow moisture to penetrate the casing, and such products do not lose any weight during processing.

6.1 NATURAL CASINGS

A large number of salami products are filled into natural casings, providing an "authentic" and "hand-made" appearance, thus increasing the value proposition for the customer. Natural casings from pigs, sheep, horses, and cattle have been used in the production of salami-type for centuries. They are highly permeable to smoke and moisture and therefore perfectly suited for the production of salami. Natural casings also display excellent coherency to the meat-mass filled into the respective casing, thus not separating from the meat-mass as occurs during drying of salami exhibiting good shrinkability. The fact that these casings are a "100% natural product" can also be used for marketing purposes, and the natural casings give a meat product a distinctly Old World appearance.

Natural casings are often made from the small intestines, which are made up of several layers. The innermost layer of the gut wall is called the mucosa and contains glands that play a role in secretion and absorption in the gut and the digestion of food. The next layer (counting from the inside of the gut to the outside) is the submucosa, which is predominantly made of connective tissue (collagen), and this layer strengthens the gut wall. The submucosa is followed by a layer of circular muscle tissue and then a layer of longitudinal muscle tissue. The outermost outer layer is the serosa, which is mainly made out of collagen and elastin, but fat is also present on the outside of this layer (see Fig. 6.1).

Beef casings used in the production of meat products originate mostly from the esophagus or weasand, small and large intestines, cecum (bungs), as well as the bladder. Weasands are usually used for large-diameter sausages and the length of a weasand is around 50–60 cm. Casings from the small intestines are commonly known as rounds or runners, whereas middles are casings originating from the large intestines. The total length of all small and large intestines together is around 20 times the length of the body of a cow. The small intestines, from which come rounds or runners, are around 35 m in length and have a diameter of 4–6 cm. Rounds are commonly flushed with water, inversed (turned

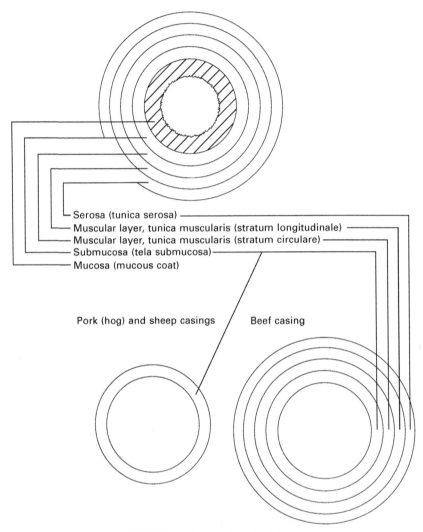

FIGURE 6.1 Cross section of an intestine.

inside-out), followed by removal of the mucosa (slimmed) as well as fat connected to the serosa. As a result, casings from the small intestines from beef contain the submucosa, both layers of muscle tissue, as well as the serosa and are therefore much thicker than pork runners, which consist of the submucosa only, with all other layers removed. Runners are sold in hanks, with one hank being 100 yards or 91.4 m in length. Bungs are made from the blind gut or cecum of cattle connecting small and large intestines. Bungs are around 70 cm long and 10–15 cm wide. Bungs are treated more or less in the same way as middles (see later). Bladders are washed, inversed, and inflated with air or salted.

Middles are separated from the ruffle (fat associated with the intestines) and flushed with water; any fat is trimmed off and they are then inversed, slimmed, and salted. Certain sections of beef middles are known as the "straight casings," which include the narrow end, wide end, and fat end, with the fat end being the mostly used part.

Pig casings have a total length of around 25 m, and parts such as the stomach, small intestines (also known as runners, rounds or hog-casings), large intestines such as cap (caecum) and middles, terminal end of large intestines (bung), as well as the bladder are used in the production of meat products. Pig rounds, from the small intestines, are around 18 m in length. To make rounds, the small intestines are separated from the ruffle, and all intestinal contents are removed either by hand or by machine. The emptied intestines are then treated with a mucus remover and slimmed (layer of mucosa removed). Machines with a set of strippers or rollers remove the mucosa, both layers of muscle tissue, and the serosa from the casing.

Finally, only the submucosa from the small intestines remains as the material to be used as casing for sausages, and it is salted. The removal of the mucosa (slimmed) used to take place in the past by hand, and to remove the mucosa as well as the two layers of muscle tissue, the casing used to be soaked in water at around 25°C overnight with the casing being inversed before soaking. Both layers would become very soft and tender and could be removed easily afterward by hand. Pig rounds, or runners, exhibit a diameter from 28–42 mm and are sometimes even larger, depending on the age and breed of the animal.

The processed pork cecum is known as cap, with the middle section of the large intestines being known as middles. The first and large sections of the large intestines of pigs are normally not used in the production of sausages. Large intestines, or middles, are separated from fat, flushed well with water, and inversed. Mucosa and serosa are removed, leaving both layers of muscle tissue as well as the submucosa. The middles are then finally salted. Pig bungs, the terminal end of the large intestine, are around 1 m in length. They are thoroughly flushed, slimmed (mucosa removed), inflated for grading, and, finally, salted. Bladders to be used as casings are emptied and excess fat is removed. Inversion takes place before they are salted or inflated by air and dried. Both hog and sheep casings are sold under the terms "selected" or "unselected" referring to a specified diameter of the casings within a hank showing a narrow and specified range, or variation, in diameter itself. Within a selected hank of casing, around 90–94% must be within the diameter specified, and selected casings are of significantly higher cost compared with unselected casings. The diameter of unselected casings varies hugely, and only around 60–70% of all casing material within a hank is within the mentioned diameter. Grading of casings toward the diameter takes place by inflating the casing with air or water to determine its diameter.

Casings from sheep are not often used in the production of salami due to their small size and diameter and are divided into rounds, caps, and straight casings.

Commonly, only rounds originating from the small intestines are used for the production of sausages such as frankfurters and salami snack products; caps are also occasionally used for salami (inversed and with the mucosa removed). The small intestines or rounds are separated from the ruffle, and the contents of the intestines are removed. The emptied casing then experiences the removal of the mucosa (slimmed) and both layers of muscular tissue as well as the serosa as only the submucosa is used as casing. The material is finally salted. Sheep casings are also sold in hanks, with one hank being 100 yards, or 91.4 m, long. The diameter of sheep casings varies commonly between 16 and 28 mm.

During salting, natural casings are heavily exposed to salt by applying around 40–50% of salt to the cleaned and prepared casing. Stored in containers at around 4°C, salted casings have an essentially limitless shelf-life as a result of the cold storage temperature and especially because of the extremely high concentration of salt, which reduces the Aw value and inhibits growth of bacteria. Before use in the manufacture of sausage, the casings are washed well to remove all salt. The thoroughly washed casings are placed in hand-temperature warm water to increase the elasticity of the casing for efficient filling and linking. Natural casings today are also available preflushed in salt brine containing around 15% salt and are already pretubed (shirred) for better efficiency during the filling process as the time-consuming process of applying the casing on to the filling pipe (stuffing horn) is dramatically shortened by the use of shirred natural casings.

Despite their "natural" appearance, natural casings carry the risk of high bacteria counts if not cleaned properly during processing of the casing, and care has to be taken that only casings with an acceptable level of total plate count find a use. Natural casings demonstrate the advantage of delayed and even drying of salami as the moisture bound within the casing acts as a "moisture buffer," thereby reducing the risk of case hardening.

Natural casings normally display a thickness of between 700 and 1300 µm, which equals 0.7–1.3 mm given that 1 µm is 1/1000th mm, or 10^{-6} m. Natural casings such as hog or sheep casings exhibit a permeability to oxygen, expressed in cm^3 per m^2 per day by atmospheric pressure ($cm^3/m^2/day\,bar$), of between 450 and 700. Collagen casing exhibit a permeability of 70–100 and are around 110–140 µm in thickness. Permeability regarding moisture (water) of natural casings, expressed in gram per m^2 per day ($g/m^2/day$), is between 1900 and 2100, with collagen casing having a permeability of around 800.

6.1.1 Collagen Casings

Collagen casings are produced from collagen extracted from the corium layer of beef hides and demonstrate more or less the same organoleptic properties as natural casings but are most often straight and uniform in shape and diameter. As such, they exhibit the disadvantage of looking "processed" and not resembling a "natural" look. On the other hand, the uniform shape and diameter support a highly efficient filling process using vacuum filling machines as well as an automated linking device. This efficiency during the filling process is very important as most collagen casings used in the production of salami are of small diameter, between 18 and 32 mm. The corium layer is the middle layer of beef hides. It is placed in a high-pH solution so that it swells and takes up a large amount of water. The swollen material is then washed, and the controlled addition of acids and application of different temperatures result in a perfectly swollen raw material. The material obtained is minced, and water, cellulose, and acids are added. All materials are cut extremely finely to obtain a homogeneous and pumpable material. Within this gel, the added cellulose functions as a glue when the gel is extruded into casings. Different extrusion techniques determine the properties of the casing, which finally passes through a solution of salt before being dried. Shirring of the casing into slugs (strands) is the next common processing step of small-diameter cellulose casings before the slugs are placed in cartons. During shirring, the straight tube of casing is compressed over a horn into a slug of casing to obtain significantly more casing material (in meters) per slug as if nonshirred by the slug being ultimately placed over the filling pipe on the filling or stuffing machine before being filled.

6.2 ARTIFICIAL CASINGS

Fibrous casings are widely used for fermented as well as cooked salami and are made out of cellulose reinforced with a fiber material such as paper for increased strength. To produce fibrous casings, paper is folded into a tube with the width of the paper determining the diameter of the tube and ultimately the diameter of the casing. The paper that is used exhibits great resistance toward heat and moisture and can be compared with paper used for the production of teabags. Once the paper has been folded into tubes, the paper tubes are run through a bath of color (in case the casing is colored) followed by several baths of acids and alkaline to "set" the casing. The last bath contains a "softener" that keeps the finished casing soft and elastic despite the fact that the casing contains only around 5–7% moisture. The moist casing is blown up to diameter and dried (at around 140–180°C), flattened, and put on rolls. The application of such high temperatures results in a casing material with a very low bacteria count and is either sold as a roll or converted further in various processing steps such as printing, shirring, cutting, or clipping.

Fibrous casings are generally soaked for around 30 minutes in lukewarm water before filling so that the material becomes soft and elastic. Printed material should be soaked for 50–60 minutes. For reasons of simplicity, there are fibrous casings on the market that are already presoaked and therefore can be filled right away without any further soaking required. Filling of fibrous casings takes place in such a way that the diameter of the casing itself is exceeded by around 7–10% once the product is filled. For example, a 60 mm-diameter casing is filled to around 65 mm. The overfilling secures a wrinkle-free product as well as making full use of the volume of the casing. Fibrous casings commonly have some kind of impregnation inside the casing that determines the cling, or adhesion of the casing, to the meat product itself. Some fibrous casings are treated in a way that the casing clings strongly to the meat products desired in raw fermented salami as the casing has to shrink with the product without becoming loose. To achieve a high cling, a layer of protein is applied on to the inside of the casing as the layer of protein present on the casings binds strongly with activated protein from the meat product filled into the casing.

On the other hand, fibrous casings can be treated on the inside with a releasing-agent that reduces cling between the casing and the meat product, securing easy peeling of the finished product. This is especially of importance for meat products that will be sliced. Some fibrous casings are not impregnated on the inside, and therefore there is some degree of cling between product and casing.

Fibrous casings to be used for the manufacture of raw fermented salami also can show a mold inhibitor as different acids are introduced into the casing. As a result, unwanted growth of mold on salami during drying is reduced. Furthermore, fibrous casings with spices and herbs on the inside of the casing are available, and once drying of the salami is completed, the casing is removed, resulting in a spice-coated end product.

The permeability of fibrous casings to oxygen, expressed in $cm^3/m^2/day$ bar, lies between 180 and 250, while the permeability to water, expressed in $g/m^2/day$, is between 2100 and 2500. Fibrous casings are around 90 μm in thickness.

The flat width (flattened tube-casing such as fibrous or plastic) is calculated following the formula:

Flat width = (caliber × pi [π])/2, with π = 3.14.

Woven textile casings are also used for high-quality products, and cotton is frequently used as the basic raw material for the production of such textile casings.

Part III

Production Technology

Chapter 7

Fermented Salami: Non–Heat Treated

The production of raw fermented salami goes back hundreds of years as it was quickly discovered that shelf stable meat products could be produced by adding salt to meat and subsequently drying the product. Fermentation of foods has been carried out for thousands of years. In earlier years, nothing, or very little, was known about parameters such as pH value and water activity but safe products were still made by relying on empirical know-how. Even today, with all the help of modern technology, know-how and practical experience are still additives highly sought after!

Salamis produced world-wide vary dramatically in quality, and the methods of production vary greatly, too. Sliceable fermented salami is a comminuted meat product consisting of meat and fat, sold raw, or in a non–heat-treated state, to the final consumer. Production of fermented salami is sometimes described as "controlled spoilage" (ie, souring and drying) of meat. Generally, salami is not a "healthy" meat product from a nutritional point of view given that the level of fat is commonly high. The level of salt added is also high, resulting in a high level of sodium within the finished product. Contrary to most other meat products, traditional salami is one of the very few meat products were no water is added during the manufacturing process. In fact, the opposite takes place and water is removed to optimize firmness, shelf-life, sliceability, and flavor.

In parts of the world such as Italy, Spain, Austria, Germany, and Hungary, the production of salami seems to be a very simple process, but in fact production of a safe, good-tasting product relies on extensive knowledge and experience gained over hundreds of years. The manufacture of salami is seen commonly as an art. Having proper equipment also plays a significant role within the production of salami given that vital changes within salami during fermentation and drying are controlled by outside parameters such as temperature, humidity, and air velocity (speed or air flow). The combination of working with raw materials, producing a raw fermented product, and having to take account of outside parameters makes the production of salami a real art form. Fermented sausages are generally very stable products, and if certain parameters are followed during their manufacture, traditional, well-loved salami can be made in a safe way. The ultimate aim when making salami is to obtain a product with a strong and stable curing color, excellent sliceability and slice coherency, typical salami flavor, and, most of all, microbiological stability.

The manufacture of salami is highly complex as product internal parameters such as Aw, pH value, Eh value (redox potential), buffer capacity of proteins, presence of NO_2/NO_3, and loss in weight change during fermentation, and these all have an effect on the color, taste, aroma, and texture of the final product. As mentioned, external parameters such as temperature, relative humidity (RH) in the fermentation room, air velocity, ripening time, application or nonapplication of smoke, and the addition of mold also play a role. Furthermore, other parameters that can vary in a salami are type and level of meat used, level of fat, initial pH and Aw value of the meat and fat materials processed, type and amount of sugar added, level of salt, spices, the presence and type of starter cultures, particle size of meat and fat, diameter of casing chosen, and whether the salami is made in the bowl-cutter or with a mincer-mixer system. It is obvious that the manufacture of raw fermented salami is a highly complex process.

7.1 SELECTION OF RAW MATERIALS

Given that sliceable raw fermented salami is by definition never heat treated, it is essential that all raw materials are of the standard required to manufacture a safe product. This refers primarily to the microbiological status of meat and fat material to be used given that high numbers of bacteria can interfere badly with fermentation and unsafe products could easily be produced. In general, to produce a safe finished product, good-quality raw materials with a low bacteria count must be used.

7.1.1 Meat

Basically any type of lean meat can be used for the production of salami, and some exotic meats such as venison, buffalo, and kangaroo have been used. Pork is in most places in the world the preferred type of meat except in countries where pork cannot be eaten due to religious or cost constraints, and salami recipes originating from countries such as Italy most often contain pork only. Boar meat has to be avoided because of its unpleasant urine-like off-flavor owing to the male hormone 5α-androsten-16-en-3-one (also known as androstenone). A major difficulty during salami manufacture is that the degree of acidification of the lean meat during fermentation is hard to foresee. This is because neither the initial pH value nor the level of sugar in the meat is ever exactly the same, and therefore, the buffer capacity of amino acids present in meat (even meat of the same type) is not always the same. Due to its naturally higher glycogen level, meat from horse and deer (venison) acidifies significantly more than the lean muscle tissue of beef and pork. The level of glycogen is higher in horse and deer meat because these animals run around as they are raised, which is not really the case any longer for cows and pigs. When processing beef and pork, the pH value within salami is reduced by around 0.15–0.3 pH unit as a result of the natural sugar (glycogen) content. Horse meat acidifies salami in a much more significant way, though, and the drop in pH value can be as large as 0.7–0.8 pH unit when all the lean meat used comes from a horse. Deer meat commonly has highly varying levels of sugar depending on the method of killing. Venison, killed in an abattoir, exhibits generally higher levels of sugar within muscle tissue than venison, shot by hunters as animals shot in the wild are not always instantly killed and during the attempt to flee high amounts of glycogen are used up.

Salami is produced either from lean muscle tissue and fat or from a combination of lean muscle tissue, fat, and fatty meat trimmings. There is no significant disadvantage in using fatty trimmings as long as the desired level of fat in the finished product is obtained. Commonly, a combination of lean meat, fatty trimmings, and fat, such as pork fat, is used.

In the production of salami, small amounts of chilled meat are used to regulate the temperature of the sausage mass overall. The chilled meat should be stored at temperatures below +4°C as temperatures below this level do not permit bacteria such as *Staphylococcus aureus* or *Salmonella* spp., two of the most significant pathogens in raw fermented salami, to grow. Carcasses should be chilled down after slaughter quickly to avoid bacteria growth overall.

The level of hygiene during the slaughtering process itself is critical in determining the bacteria count of the meat. If intestines are opened and the gut contents come in contact with meat on the carcass, those affected areas should not be washed down with water as the bacteria would be distributed over the carcass to a greater extent. Contaminated areas should be cut free by removing the area with a knife and then no further contamination of the carcass can take place.

Lean meat is not processed when frozen because fully frozen meat cannot be cut properly in the bowl-cutter and cannot be minced to the desired particle size as it would break mincer plates. Lean meat used for the manufacture of salami is processed either in a semifrozen (tempered from frozen) state (see Section 3.4 under Chapter 3: Definitions) or predried state and bacteria growth has to be kept to an absolute minimum during tempering. The surface of tempered meat is particularly prone to rapid bacterial growth during further thawing. Tempered or semifrozen meat displays a temperature of between −10 and −5°C and acts as a cooling-agent for the fat present within the sausage mass during cutting in the bowl-cutter or during mincing. A frozen fat surface during comminution is maintained and, as a result, fat is cleanly cut in the bowl-cutter or mincing reducing the risk of fat-smearing, which is very beneficial during the drying-phase of salami. Freezer-burned meat (see Section 3.5 under Chapter 3: Definitions) frequently displays discoloration on the surface and should be avoided. Even the leanest meat contains some fat and meat, stored by −18 to −20°C, should not have been placed in the freezer for longer than 8–12 months. Some degree of rancidity develops over time, which has a negative impact on the flavor in the finished product.

Meat used for the production of salami must not contain glands, sinews, or blood clots. Glands commonly exhibit a high number of bacteria, which can interfere with fermentation, and blood clots are also very susceptible to spoilage given that they commonly have a high bacterial count as well. Sinews should be removed given that fermented salami is never exposed to heat treatment and sinews present within the product always remain tough.

The bacteria count of meat to be processed should be as low as possible and numbers between 10^2 and 10^3/g are seen as optimum. Numbers above 10^5/g of meat should be seen as the maximum, and the number of Enterobacteriaceae overall should be less than 10^4/g. Any unwanted bacteria introduced into the raw salami mass compete with starter cultures (if used) for food during fermentation and can cause the fermentation to go from control.

The pH value of meat to be processed should be below 5.8–5.9: WHC is not a desired quality as the salami should decrease in weight during production and high pH values in meat support WHC. It is disadvantageous to use DFD meat (see Section 3.1 under Chapter 3: Definitions) as DFD meat has undergone incomplete rigor mortis and is microbiologically not as stable as fully acidified meat, which underwent complete rigor mortis. The high pH value in DFD meat also correlates with a high WHC. A high WHC is not desired in the production of salami given that the aim is to remove water from the product. An increased pH value also often correlates with a slightly increased bacteria count given that elevated pH values favor bacteria growth as well as enzyme activity. Using DFD meat can also cause difficulties during the acidification of the salami, and the pH value could, as a result, not decline to the desired level, thereby putting the safety of the product in question. DFD meat presents a real problem in salami products containing a high level of beef, or in all-beef products, if such meat exhibits DFD character. In salami containing levels of beef between 10% and 20%, DFD is not a "critical" problem as other meats, mostly pork, secure the required acidification.

High pH values in meat are also not beneficial for the development of the curing color in the finished product as less undissociated nitrous acid is present and a reduced amount of NO is obtained (see Section 5.3 under Chapter 5: Color in Cured Meat Products and Fresh Meat). Beef meat from cows rather than bulls is often preferred as it demonstrates a darker color, which is an advantage in producing salami. Cow meat is preferred for several reasons. As well as having a higher level of myoglobin compared with the meat of young bulls, which is beneficial for a strong curing color, cow meat generally exhibits a slightly higher level of glycogen compared with meat from bulls, therefore acidifying in a more significant way. It is less expensive than other beef meat and exhibits a lower water activity compared with muscle tissue from younger cattle, which shortens the drying process.

PSE meat (see Section 3.1 under Chapter 3: Definitions) is of no disadvantage within the production of salami as a reduced WHC aids the loss in weight during drying. However, if the level of PSE meat within a batch of salami containing high levels of pork goes beyond 20–25%, a loss is color can be observed. Small batches of salami made with pork meat from only one pig can be problematic if the pig has PSE-character meat. With larger batches, a mix of meat originating from several different pigs is commonly processed and the risk of pale color in the finished product due to high levels of PSE meat is minimized.

Meat materials used for the production of salami must be free of any residual antibiotics given that those substances interfere badly with fermentation and faulty products will be obtained. Pork neck and pork shoulder are prone to contain pus, and therefore if meat from the neck and shoulder is used, several cuts should be made with the knife into the meat before freezing to check that there is no pus within muscle tissue. This is especially the case if those cuts are from sows, as injections, a cause of pus formation, are commonly performed by veterinarians in the neck of sows.

When pork is used, sow meat (old female) is a preferred choice of material as it has a strong red color and exhibits a slightly reduced Aw compared with pork meat from young pigs. A decreased Aw shortens the period of drying, which can be an economic benefit. Generally, meat from all different cuts from sow, such as shoulder, belly, leg, and the loin, can be used for sliceable salami. From the belly, the soft areas along the nipples as well as the flank part in front of the leg should be avoided for the manufacture of salami given that those areas contain fat with a high level of unsaturated fatty acids. That fat is very soft in consistency and the fat is easily smeared during cutting or mixing, which has a negative impact on drying afterward. Fat of such soft consistency can be perfectly used for spreadable raw sausage, cooked sausage such as frankfurters, and liver sausages.

Cuts of fresh chilled meat from the carcass of sows such as shoulder, leg, neck, and belly are frequently preconditioned, or predried, before use or being frozen and are hung in a chiller at around −3 to 0°C under high air velocity for 2–3 days to remove a fair amount of moisture, as such reducing the Aw value in the meat, even before the material is introduced in the salami mass. Predrying of meat only makes commercial sense if the end product is released for sale on reaching a specified Aw value and release of the end product is not based on a specified weight loss in percentage. The weight loss in percentage is generally based on the loss in weight during fermentation and drying of the salami after being filled into the respective casing and does not account for any weight loss before filling of the salami. As such, the weight loss during predrying of the meat would be an economical loss. Predried and semifrozen material is also often used to raise the overall temperature of the salami mass to around 0°C when all other meat and fat materials used are semifrozen.

Predried meat is also applied to raise the temperature of the salami mass because if all the meat materials were semifrozen, the temperature of the sausage mass would be too low for proper fermentation afterward and the filling process would be negatively affected as well. Another way of using the predried material showing a temperature between −2 and 0°C is in salami in which all meat and fat materials are minced to obtain the desired particle size as the predried meat displays the perfect temperature for further processing.

In salamis cut in the bowl cuter to a particle size between 2 and 4 mm around 10–20% of minced chilled (2–3 mm) meat is added at the end of the cutting process to the cut salami mass to raise the temperature of the sausage mass. During mincing of the meat a temperature below +4°C must be maintained to avoid growth of bacteria. The addition of around 10–20% of chilled minced meat raises the temperature of the sausage mass to around −2 to 0°C, which is perfect for filling despite using fully frozen fat and tempered meat within the recipe to around 80%.

Freeze-dried meat can be used to lower the Aw right from the beginning of the fermentation process. The maximum amount of freeze-dried meat added to a batch of salami is around 10% of the total amount of meat given that the addition of excess levels would reduce the Aw below 0.95 at the beginning of fermentation and starter cultures such as *Lactobacillus* spp. would have insufficient free water to ferment added sugars into lactic acid. If this were the case, acidification within the product would not occur properly and very poor curing color, no slice-coherency and a microbiological unstable product would be obtained (see Section 7.4).

Soft MDM meat (see Section 3.2 under Chapter 3: Definitions) is also used in the production of salami and is predominantly introduced into minced, coarse salami products. Materials that are high in connective tissue such as shank meat or other trimmings put through a soft MDM machine and such material, generally exhibiting a particle size of 4–6 mm, can be used perfectly well for salami. Material to be put through the separation machine is placed in an area with a temperature of around −2 to 0°C overnight and is minced afterward with a 13- to 20-mm blade before being processed in the soft MDM machine. The low temperatures of the meat material supports a clean mincing process of collagen-rich materials and the MDM machine can separate collagen from lean meat more effectively afterward as well.

The nonmeat part originating from the separation process is mostly connective tissue, which can be perfectly used in cooked sausages like frankfurters or even turned into a material used in salami. To make it usable for salami, the connective tissue is placed into waterproof casing and cooked for around 3–4 hours at around 90°C. During this thermal treatment, a high amount of connective tissue, and more precisely, collagen, gets softened and some even turns into gelatin. Following cooking, the casing is removed from the still-warm product and the cooked material is placed into trays, which are stored in the freezer. The frozen material can be cut with a frozen meat cutter and subsequently turned into a dust-like material by cutting the fully frozen in the bowl-cutter with sharp knives. The dust-like material can be introduced at levels of around 1–2% to salami without changing the product overall.

Salami produced from chicken meat mainly contains thigh meat with the skin on as thigh meat results in a significantly stronger curing color compared with chicken breast meat. Thigh meat is most often also considerably less expensive than breast meat. Care has to be taken that the meat is free of bone particles and the skin must not be rancid given that chicken skin contains a fair amount of fat and chicken fat is prone to rancidity.

Mechanically deboned meat from chicken (chicken MDM), also known as hard MDM, is not used in raw fermented sliceable salami as it often has a high bacteria count as well as large surface area. Hard MDM also frequently contains up to 15% fat and as it has such a high level of fat, present in a large surface area, it is very prone to rancidity. Most countries in the world do not permit the use of hard MDM in raw fermented salami by law.

7.1.2 Fat

Fat is cheaper than lean meat and therefore 25–35% of fat is commonly added to salami raw materials. During drying, a process in, which water is predominantly lost from lean meat, the level of fat in the fully dried salami rises to 40–50%, depending on the degree of drying loss as well as the level of fat introduced in first place. Besides the economical factor, fat is also acts as a carrier for flavor because countless flavors, or flavoring substances, are fat soluble. Fat or very fatty trimmings are processed in a frozen state as smearing of fat during processes such as cutting or mincing has to be avoided.

Pork fat is the most commonly used type of fat for salami manufacture as it demonstrates organoleptic properties far superior to beef or chicken fat. Pork back and neck fat is the hardest fat within a pig and is therefore the first choice when producing salami due to its low content of unsaturated fatty acids such as linoleic and linolenic acid compared with fat from other parts of the pig. Fat from neck and loin (back fat) has the lowest number of unsaturated fatty acids and therefore does not smear as easily during cutting or mincing compared with soft fat coming from shoulder or leg. Soft-textured fat, such as that from the shoulder or leg, has higher levels of unsaturated fatty acids, which reduce the firmness and melting point of the fat. Different levels of hardness within the same piece of fat can be seen. Fat coming from the loin, for example, consists of several layers of fat, with the layer directly underneath the skin being the softest and the layers, closer to the inside of the carcass, being of harder consistency. The level of saturated fatty acids within the different layers of fat increases from the outside inward and subcutaneous fat has the highest number of unsaturated fatty acids. Fat within pork belly is also of much harder consistency than soft fat from shoulder or leg. Generally, the hardness of fat is also influenced by the content of connective tissue within fat and fat tissue from older pigs is firmer overall.

Fat materials used for salami must not be rancid and pork fat should be white, or whitish. As with lean meat, the bacteria count should be as low as possible (around 10^3/g) and no remains of pork skin must be present on the fat. The deskinning machine has to be adjusted so that during the separation of skin from fat, no skin remains on the fat itself but as little as possible of fat remains on the skin. Beef fat as such is hardly used in salami. Beef fat is commonly found in fatty beef trimmings though, and so can be found in this form in salami. Pure-beef salami is most often produced from fatty beef brisket, especially in countries where people do not eat pork due to religious belief.

Different levels of fat within the salami mass influence the pH value to a degree since fat has a higher pH value than muscle tissue. High levels of fat increase the pH value of the sausage mass slightly but are of no, or very little, significance during fermentation. The impact of high levels of fat on the Aw of salami, though, is much greater given that fat contains only around 15% water while lean meat contains around 75% water. A recipe high in fat has a lower Aw than a recipe "low" in fat at the beginning of the fermentation process. As a result, the RH has to be higher during the beginning of fermentation to avoid case hardening.

Any fat used must not be rancid given that over the prolonged period of fermentation and subsequent drying the level of rancidity would increase exponentially. The type of feed given to pigs has a major impact on the composition of the fatty acids finally present in fat. Feeding of oily substances results in fat demonstrating high levels of unsaturated fatty acids thus softening the fat material. Increased levels of unsaturated fatty acids also speed up rancidity and periodic checks of fat quality such as the iodine index (measures the level of unsaturated fat) or the acid index (indicator for freshness) are practiced.

Loin and neck fat from sows is commonly preconditioned for 2–3 days at −4 to −2°C under high air velocity before freezing. This reduces the level of moisture within the material so that less water needs to be removed during drying of the salami. During this process the water content of fat is reduced by around 5% and the melting point is increased as the same time (altered fat structure).

Beef fat, coming from cuts such as brisket with high levels of saturated fatty acids, can be used within salami if it is frozen and then cut in the bowl-cutter under high speed to obtain a dust-like material. This dust-like material can be introduced into almost any type of salami at around 1% without changing any product characteristics. The use of beef fat in salami is economically beneficial as beef fat would otherwise generally be discarded and has no value. In most cases, manufacturers even have to pay for the fat material to be picked up for disposal. It goes without saying that beef fat used must show an excellent microbiological status.

7.1.2.1 Fat and Meat Replacers in Salami

Lean meat is the most costly of all the raw materials used within salami. A large amount of salami is sold at a low price (to be used, for example, as pizza topping) and meat replacement is one of the very few possibilities to lower the cost of the raw materials of salami.

A gel consisting of soy isolate and iced water in a ratio of 1:3 can be used as a meat replacer in salami production. The soy isolate used for such an application must demonstrate high gelling properties. Iced water is placed in the bowl-cutter and soy is added to it while cutting under medium-fast speed and after only several minutes of cutting a spongy-firm and dry mass is obtained. Colors such as fermented rice, carmine, and/or caramel are needed to simulate a meat-like color within the gel. Those colors are introduced into the water first and fully dispersed before soy is added. The color of the gel has to be dark given that the color of meat within salami darkens during drying while the color of the gel remains unchanged. The color of the gel made therefore has to be very similar to the color of the meat within the salami after drying is completed. The spongy-firm gel is removed from the cutter, placed into trays, and stored under chilled conditions overnight. A firm, spongy, and sliceable gel will be seen next day, which can be introduced into the salami mass.

It is common to substitute between 5% and 30% of the total lean meat content and the gel is to be treated as lean meat during the process. During fermentation, acidification also stabilizes the granules of gel against microbiological spoilage.

Replacement of fat within salami is sometimes desired even though it is well known that salami is not a lean or low-fat product as such. Fat can be replaced in salami using a soy isolate–oil emulsion. The emulsion can be made from 1 part soy isolate, 2.8 parts iced water, and 2 parts oil. The iced water is placed in the bowl-cutter and high gelling soy isolate is added into it while cutting under medium-fast speed. Once all water is fully absorbed by the soy protein and a dry mass is obtained, oil is gradually added. Introducing oil lightens up the color of the emulsion and once the oil is fully emulsified the mass is removed from the cutter and placed into the chiller overnight to firm up.

Another method of replacing fat in a salami is to produce a fully cooked emulsified sausage. A white-colored sausage can be produced using chicken meat, chicken skin emulsion, and starch. The recipe for this type of cooked sausage, used finally as a fat-replacer in salami, contains around 60–70% fatty chicken meat, 15–20% chicken skin emulsion, and around 10–15% water/ice. Additives introduced are phosphates, salt, nitrite, and starch with starch being added at a level of between 4% and 6% based on the total mass. The aim is to obtain a spongy-firm and white-colored emulsion, which is filled into a waterproof casing and cooked to 72–74°C. The cooked and chilled product can be introduced into salami for the purpose of imitating fat. Especially in chicken salami, such a white-colored cooked sausage mass finds its use and provides a nice contrast to chicken thigh meat.

7.2 SELECTION OF ADDITIVES

Salt is the oldest additive in the world and has been used in the preparation of food for thousands of years. In salami, salt is applied for several reasons. First, salt is the first hurdle against bacterial growth and is applied at levels of between 26 and 30 g/kg of sausage mass. Levels of salt below 25 g/kg of salami are not recommended given that levels above 25 g/kg are significantly more effective in inhibiting the growth of bacteria by reducing the Aw within the sausage mass.

Inclusion-levels below 25 g/kg are insufficient for development of an effective hurdle. The addition of salt at recommended levels lowers the initial Aw within the salami mass to around 0.96–0.97. This is also the reason why salt-reduced salami present a food safety risk given that a reduced level of salt has a weaker impact against bacteria growth of "unwanted" bacteria during the initial stages of the fermentation process. Second, salt is a flavor enhancer and meat products do not taste good without salt. Third, salt aids the activation of protein, which is necessary to obtain slice coherency in the finished product (the salt contributes to the formation of a sol during cutting of mixing of the salami mass). The addition of salt also reduces the temperature of the salami mass by around 1–2°C. The relatively high level of salt present within the sausage mass reduces the freezing point of water within lean meat to around −4°C, permitting subzero temperatures within the sausage mass. Therefore, smearing of fat during the cutting and filling process is avoided.

The majority of salami, especially all fast and medium-fast fermented products, are produced nowadays with added nitrite (NO_2). Nitrite is another important hurdle against microbiological spoilage especially during the initial stage of fermentation. It is introduced into the sausage mass commonly as sodium nitrite at greatly varying levels depending on the maximum amount permitted in the finished product as per food law in the respective country. In some countries, such as the United States, the level of nitrite added to salami is based on the time when the salami is produced and not when the salami has completed its drying process and is released for sale. Inclusion levels between 150 and 500 ppm/kg of raw, unfermented sausage are possible based on varying food laws around the world. Around 130 ppm of nitrite/kg of sausage mass suppresses the growth of Enterobacteriaceae, such as *Salmonella* spp., and other gram-negative bacteria. Nitrite plays a major role in controlling bacterial growth in salami manufacture, in particular in fast fermented salami, when the temperatures is raised to between 26 and 30°C during fermentation. The effect of nitrite as a hurdle against microbiological spoilage is much greater in salami than in other meat products such as cooked ham or cooked sausage. This is because nitrite is more effective at low pH values and acidification occurs during fermentation of salami. In meat products such as cooked ham and cooked sausage, the pH value is raised due to the addition of alkaline phosphates and the nitrite is not as effective.

Significant amounts of nitrite added to the sausage mass in the first place are converted into curing color, curing flavor, and other substances within 3–4 days after production as nitrite reacts in countless different ways under the impact of increased temperatures during fermentation. When additives such as GDL or ascorbic acids are also introduced, the conversion of nitrite into numerous other substances is speeded up even more. Nitrite is also the primary agent for development of the desired curing color in a salami and it also contributes to curing flavor. Last, nitrite also acts as an antioxidant. The antioxidative effect of nitrite comes from the oxidation of nitrite to nitrate. Hence, NO obtained from nitrite binds to the iron core of myoglobin and hemoglobin thus reducing the release of iron, which if not bound would promote oxidation of fat. The binding of NO to myoglobin or hemoglobin instead of oxygen is due to the fact that NO has a much stronger affinity to those substances compared with oxygen. Nitrite also has the ability to chelate free iron, thus also deactivating this pro-oxidative material. There is suspicion that small amounts of nitrosamines as well as secondary and tertiary amines can form from nitrite under the slightly sour conditions, which exist within salami as a result of acidifying the product when salami is exposed to high temperatures such as being used as pizza topping.

Nitrate (NO_3), rather than nitrite, mostly used in form of potassium nitrate, is added during production of slow-fermented salami. Nitrate does not have a significant impact against bacterial growth. To contribute toward development of curing color, nitrate has to be reduced first to nitrite before it can bind to myoglobin to form nitrosomyoglobin (see Section 5.3 under Chapter 5: Color in Cured Meat Products and Fresh Meat).

Antioxidants used in salami are ascorbic acid, ascorbate, erythorbate, and occasionally tocopherol. Ascorbic acid/ascorbate acts as a direct but also secondary antioxidant by donating H^+ to the sausage, which deactivates free radicals. In a secondary function it supports antioxidative substances such as phenols (which act as antioxidants by absorbing free radicals into their ring-shaped structure), by donating H^+ to phenol radicals for their reduction to phenol. Tocopherols in cured meat products are able to reduce the number of nitrosamines formed and given that tocopherol is fat soluble, it is an excellent antioxidant in slow-fermented salami in terms of retarding rancidity. Reducing the redox-potential (Eh value) within the sausage mass at the beginning of fermentation through the impact of antioxidants is also advantageous for the lactic acid bacteria. This is because competitive microflora, such as aerobic bacteria like *Pseudomonas* spp., are less able to survive when levels of oxygen are reduced. A reduced Eh value also makes NO_2 more effective as a hurdle. On rare occasions, rosemary extract is used as an antioxidant as well.

Ascorbic acid and ascorbate (erythorbate) also act as color enhancers and are introduced at 0.5–0.7 g/kg of sausage mass. Excess levels of ascorbic acid/ascorbate can favor the growth of unwanted bacteria and levels of 0.5–0.7 g/kg are fully sufficient. A slightly larger quantity of ascorbate compared with ascorbic acid has to be added to fulfill the same function (see Section 5.4 under Chapter 5: Color in Cured Meat Products and Fresh Meat) as a color enhancer. In the manufacture of fermented salami today, ascorbic acid is commonly used in very fast-fermented products. In most cases, though, ascorbate, or erythorbate, is applied as it stabilizes the curing color during the product's long drying period. When ascorbic acid is used, it must neither come into direct contact with nitrite within a premix nor be added at the same time as the nitrite to the sausage mass. Ascorbic acid, if used, should be added at the very beginning to the sausage mass. Salt and nitrite are commonly added evenly at the end of the cutting or mixing process. Rosemary extract also is applied nowadays as an antioxidant especially in long-dried salami to avoid rancidity. The level of such material applied is based on the concentration of carnosic acid within the material itself and usually amounts between 0.2 and 0.5 g/kg of salami are applied.

Spices are introduced into salami according to taste, and care has to be taken as some natural spices may contain high numbers of bacteria. Treated spices, with a significantly reduced number of bacteria, should be the preferred choice. Most salamis contain pepper and garlic as their basic flavor, while spices such as chili and fennel are also frequently applied. Garlic also has the ability to inhibit growth of bacteria. The amount of garlic that would have to be added, though, for it to have a bacteriostatic effect would not be tolerated by the consumer.

Other spices commonly introduced are coriander, mace, nutmeg, paprika, and chili. Spicy-hot materials such as chili powder or cayenne pepper must not come in direct content with nitrite because the acids present in such spicy materials react with nitrite causing the loss of nitrite and the technological functions needed from the additions of nitrite are lost. Mustard-powder seed can be added at 1–2%. It reduces the Aw of the sausage mass, contributes positively to the flavor, and delays the onset of rancidity. Mustard powder is made from enzyme-deactivated mustard seed and does not have a mustard flavor given that the enzyme responsible for creating the mustard flavor is destroyed during heat treatment. Mustard seed also immobilizes free water as it contains around 25% protein and contributes positively to the firmness of salami. Despite its high level of protein, mustard seed powder is generally declared as a spice. Colors such as chochenille are also permitted in some countries but are commonly not applied with salami except in countries such as France and China, where chochenille is frequently applied.

Smoke is an additive commonly applied within the production of salami as it aids preservation (phenols), hardens the surface, acts as an antioxidant, limits growth of unwanted mold, and has a favorable impact on color, flavor, and taste (see Section 4.15 under Chapter: Additives). Liquid smoke can be applied at 0.05–0.2 g per 1 kg directly into the sausage mass to impart a smoke flavor to the end product without the product being actually smoked. Some countries permit the use of sour phosphates, which promote easier filling or better slipping of the sausage mass during filling, reducing the risk of fat smearing.

Long cellulosic fibers originating from materials such as wheat, carrot, potato, citrus, or soy are added to fermented salami to reduce the Aw value of the product as well as to increase the speed of drying. The underlying theory for an increased speed of drying is that the fibers introduced create a secondary pipeline or drainage system allowing a faster removal of water from the product without obtaining case hardening and drying time can be reduced by as much as 25%. The level of fibers added per kilogram is salami mass varies between 3.0 and 10.0 g. Case hardening, or a dark drying ring, is obtained in the outer layer of a salami when moisture is removed faster from the surface of the salami as moisture penetrates, or diffuses, from the core of the salami toward the surface. As a result, the "stream of water" from the core toward the surface "breaks" and a hard ring is obtained. Obtaining case hardening slows down further drying significantly and does not look visually attractive. When severe case hardening is obtained in the early stages of fermentation, and little moisture is removed from the product, "unwanted" bacteria naturally present in meat can overthrow the activity of starter cultures, leading to a microbiologically spoiled product.

7.2.1 Selection of Acidification Additives

Most salami, except for slow-fermented salami, acidifies to a greater or lesser extent during fermentation. Certain additives are of great help in achieving a secure and sufficient acidification, which has an impact on color, sliceability, texture, flavor, and the microbiological stability of the product.

7.2.1.1 Chemical Acidulants (GDL, Citric Acid)

Glucono-δ-lactone (GDL) (see Section 4.18.1 under Chapter: Additives) is the ester of gluconic acid and hydrolyzes to gluconic acid once introduced into salami due to the presence of water, which originates from the meat itself. The addition rate of GDL to fermented salami varies between 3 and 12 g/kg of product. Encapsulated GDL can be applied to delay acidification, but is generally not applied as the primary purpose of adding GDL is to achieve fast and secure acidification inside the salami. On an average basis, around 8–10 g of GDL are added per 1 kg of salami when acidification is caused entirely by GDL, reducing the pH value from around 5.6 to 4.6–4.8 during fermentation. GDL not only produces gluconic acid but also causes the formation of small amounts of acetic acid. Excess levels of GDL can cause bitter flavor and the pH value within salami is reduced by around 0.1 pH units (for example, from 5.6 to 5.5) through the addition of 1 g (or 0.1%) of GDL per 1 kg of sausage mass. High levels of GDL also can cause bluish spots on the casing, which disappear during the processing steps of fermentation as well as subsequent drying. This phenomenon has as yet not been fully explained.

The transformation of GDL into gluconic acid depends greatly on the temperature of the sausage during the first 24–76 hours of fermentation. Higher temperatures during this period of time increases the rate of transformation given that any chemical reaction occurs at a faster rate at elevated temperatures. As a rule of thumb, an increase in temperature of 10°C speeds up a chemical reaction by 100% or, in other words, the same reaction takes place in half the time. Since the formation of gluconic acid from GDL in the presence of water is a chemical reaction, the formation of gluconic acid cannot be stopped as long as there is GDL and free water inside the salami (the formation could in theory be stopped by freezing of the product, but this is not an option). Even a reduction in temperature does not stop the formation of further gluconic acid as reduced temperatures only slow the process down. High levels of GDL promote the growth of peroxide-forming *Lactobacillus* spp., which can lead to rancidity and poor curing color. The acidic flavor from gluconic acid, from GDL, is less favorable compared with lactic acid originating from starter cultures fermenting sugars given that lactic acid displays a much more pleasant and typical salami acidic flavor.

7.2.1.1.1 Citric and Lactic Acid

One gram of citric acid reduces the pH per 1 kg of salami mass by around 0.2–0.3 pH unit. The acidification capacity of citric acid is therefore around 2–3 times greater than GDL. However, citric acid is rarely used on its owns because the rapid drop in pH value allows insufficient time for the development of curing color as well as curing flavor and contributes to a sour taste in the product. Citric acid is commonly applied in combination with GDL or in an encapsulated form in cooked salami to cause a "delayed" acidification when a salami product is heat treated. The coating materials applied for coating citric acid melt at around 50–55°C, and once those temperatures are reached within the cooked salami, acid is slowly released.

As with citric acid, lactic acid also causes a rapid decline in pH value if applied in a nonencapsulated form.

When the drop in pH value in salami is rapid until the isoelectric point (IEP) at a pH of 5.2 is reached, its moisture content remains high. If the sol, which is obtained in the salami during cutting or mixing, is transformed quickly into a gel at the IEP, there is little time for water to be removed during the period when the pH value declines from around 5.7 down to the IEP. The period in which the pH value drops from its starting point down to the IEP at 5.2 is the perfect time to remove high amounts of moisture from a salami. Therefore, the level of moisture in the product will be too high, necessitating prolonged drying to obtain the correct weight in the finished product. If the IEP is reached and passed through quickly, the situation is even worse as the water present within proteins becomes bound due to their rapid denaturation. Only small amounts of moisture are lost until a pH of 5.2 is obtained in case the pH drops at a rapid rate and drying afterward is slowed down as well as water becomes entrapped in the rapidly denatured protein structures. This is not of economic benefit.

7.2.1.2 Sugars Used for Acidification and Flavor in Salami

Different types of sugars are commonly used in the production of salami. Their primary function is to provide food for starter cultures, and therefore they are fermented into other substances, preferably lactic acid. Added sugar also smoothes the flavor of salami and reduces the Aw in the product. Therefore, it is a minor hurdle against microbiological spoilage.

The decline in pH value in the product depends largely on the type and amount of sugar introduced into salami in first place. Elevated levels of sugar leads generally to a stronger acidification and therefore lower pH values. To be fermented into lactic acid, sugars such as sucrose, lactose, and maltose must be broken down first into monosaccharide.

Glucose, on the other hand, can be fermented directly into lactic acid and is therefore by far the most often applied form of sugar in fermented salami. The production and ratio of D- and L-lactic acid in the salami depends on the species of lactic acid chosen as being the starter culture. Sucrose is the second fastest fermentable sugar. Maltose and lactose require a considerably longer period of time for the glycosidic bonds in their molecules to be broken until fermentable monosaccharide are produced. In essence, all lactic acid bacteria (LAB) can ferment glucose into lactic acid. Sucrose can be fermented by around 85% of LAB, maltose by around 70% of LAB and lactose by only around 55%. Only around 30% of lactic acid bacteria ferment galactose into lactic acid.

Sugars, which are not directly or only partially fermented (eg, lactose, maltose, and galactose in some products), play a role in color and flavor development.

Generally, 1 g (or 0.1%) of dextrose added per 1 kg of salami lowers the pH by 0.1 pH unit, which is equal to reducing the pH by 1 unit when adding 1%, or 10 g, of dextrose per 1 kg of salami. Therefore, 8–10 g of dextrose reduces the pH in salami from around 5.7 to around 4.6–4.8, which is frequently the final pH desired. Similar declines in pH value can be achieved by adding 7 g of dextrose or 2–4 g of lactose. Sugar does not contribute to the formation of curing color directly, but the fermentation of sugars into acids reduces the pH value within the salami and higher levels of undissociated nitrous acid (HNO_2) are obtained as a result. Increased levels of undissociated nitrous acid lead to a stronger curing color (see Section 5.3 under Chapter 5: Color in Cured Meat Products and Fresh Meat). Sugar can also be used as food by bacteria such as *Micrococcus* spp., which reduce nitrate to nitrite, also helping to develop a good curing color. Maltodextrines with a DE value of 20–35 are also used in salami given that having different DE values present leads to a partial formation of lactic acid, which is desired. When only maltodextrines are applied for the purpose of acidification, significantly higher levels have to be introduced than if glucose were added as maltodextrines have limited acidification capacity. This is rarely the case, though, given that dextrose is such a low-cost material. Sugars or carbohydrates such as maltodextrine, sucrose, and lactose do not only contribute to the formation of lactic acid but also contribute to formation of a good flavor in salami. Processing parameters such as different speeds of acidification, the degree or level of acidification, and contribution to flavor overall make it feasible to introduce blends of sugars into salami rather than a single type of sugar.

7.2.2 Starter Cultures

Salami factories very often develop their own house flora, or factory-specific monoculture, over time. When using a house flora, to produce fresh salami, though, fermentation is not reliable or consistent. Therefore, for many years, salamis have been inoculated with a concentrated and selected mix of bacteria, or starter culture, to begin fermentation. The use of starter cultures means that the proper type of bacteria in the amount required is added to the sausage mass to ensure efficient and safe fermentation. Before starter cultures were used, a back-inoculum used to be applied to begin fermentation (back-slopping). The back-inoculum was a leftover from a previous fermented batch of salami and used as starter culture for a new batch to increase the number of LAB present in the raw materials. This method of starting fermentation is not practiced anymore in modern production facilities because when a back-inoculum is used, all the bacteria present in the inoculum, including the unwanted ones such as heterofermentative *Lactobacillus*, are introduced into the freshly prepared batch of salami. When high numbers of unwanted bacteria are present in the inoculum, they automatically contaminate the new batch of product. A new batch of salami produced with a back-inoculum is also not fully traceable as a nonstandardized material has been used.

Starter cultures are selected bacteria and are added to salami for their positive contribution to acidification (and therefore microbiological stability of the product), color, and flavor. The number of starter cultures added to the salami mass has to be at least 10^7/g of mass, and this figure shows clearly that raw fermented salami is a living material and millions of bacteria are present within every gram of product. Starter cultures must not harmful to human health, must be tolerant against high levels of salt as well as nitrite, and must work at low temperatures as the salami mass has a temperature around 0°C after being filled into the respective casing.

Starter cultures are sold frozen, freeze-dried, or in liquid form and most are nonproteolytic and nonlipolytic. Liquid nitrogen or CO_2 is often used to freeze starter cultures as nitrogen has a temperature of around −195°C and CO_2 has a temperature of around −75°C. Starter cultures should also not produce CO_2 or H_2O_2 and should grow optimally at temperatures between 22 and 38°C. When freeze-drying starter cultures, water is first extracted using a vacuum. While stored frozen, the cultures are dormant and once they have sufficient water again, coming from meat, they regain their original activity. Freeze-dried cultures can be mixed for around 15–30 minutes in cold water before being introduced into the salami mass to make them faster acting but this is not essential. Premixing with water makes supports an even introduction into the sausage mass as the amount of starter cultures used per batch of salami, without water, is rather small. If premixed with water, demineralized water should not be used as cells would die by bursting in this type of water. The bacteria can, as said earlier, be premixed with water for around 15–30 minutes before use, rather than for a period of hours, as some degree of unwanted activity or fermentation would take place during this period of time. Types of sugar, such as glucose, are predominantly the carriers used in commercial starter cultures, and so if the culture was soaked for too long, it would start to ferment the sugars present.

Around 10^{10}–10^{12} of bacteria are added per 100 kg of salami and these 10^{12} bacteria only weigh around 1 g on their own. Therefore, carriers are used as they make starter cultures easier to handle. The number of foreign bacteria within starter cultures should not exceed 10^3/g of culture and certain bacteria such as *Clostridia* spp. and *Salmonella* spp. must not be present at all.

Starter cultures are introduced into the sausage mass at the beginning of the cutting or mixing process and they must be evenly distributed. Uneven distribution results in nonuniform acidification and therefore faulty products.

Nowadays, a mix (or blend) of starter cultures is commonly added to fulfill more than one criterion, so that the culture added is active at a broad range of temperatures and levels of humidity during different stages of fermentation and drying.

Freeze-dried or frozen starter cultures should be stored at −18 to −25°C until use and should not be premixed with other additives such as sugars, GDL, ascorbate, spices, and salt before use. If premixed with other additives 1 or 2 days before being added to the sausage mass, some moisture is always present and as soon as the starter cultures have access to humidity, temperature, and some food (such as sugars), fermentation would begin and faulty salami would be the result. The acid produced by a starter culture premixed with other additives, even if the premix were stored in a chiller, would be sufficient to react with nitrite and the salami would be discolored as a result. The acid produced within the premix would subsequently be missing inside the salami and insufficient acidification will take place leading to microbiological spoilage of the product and a risk of food poisoning. These vital points explain why complete blends or premixes of additives containing starter cultures for the production of salami do not exist.

Members of the genus *Lactobacillus*, *Staphylococcus*, *Pediococcus*, and *Micrococcus* are important in salami starter cultures. Microorganisms belonging to the family Lactobacillacea are the most significant in starter cultures in general and bacteria such as *Lactobacillus (Lb.) plantarum*, *Lb. casei*, *Lb. acidophilus*, *Lb. brevis*, *Lb. sake*, *Lb. curvatus*, *Lb. lactis*, and *Lb. fermenti* are often used. Commonly, *Lb. plantarum*, *Lb. sake*, *Lb. lactis*, and *Lb. curvatus* are exploited in salami and lactic acid is obtained when lactobacilli ferment glucose via the Embden-Meyerhof pathway (glycolysis). LAB are added at levels of 10^6–10^7/g of salami and homofermenative species are chosen. *Lb. plantarum* ferments a wide range of sugars, while *Lb. curvatus* ferments mainly sucrose and lactose. *Lb. plantarum* also neither forms gas from glucose nor reduces nitrate to nitrite. Some strains of *Lb. sake* and *Lb. curvatus* produce H_2O_2 in the presence of oxygen. LAB in salami starter cultures should be preferably homofermentative (homolactic) and should ferment different sugars predominantly (preferably at a level of more than 90%), into lactic acid. Heterofermentative Lb., on the other hand, only ferment sugars partially into lactic acid but also into significant amounts of acetic acid, ethanol, CO_2, and H_2O_2, all substances that are not desirable. If gases such as CO_2 form, pores are produced in the salami or vacuum packages are blown up by the gas. *Lactobacillus* spp. are mostly anaerobic bacteria. They cannot use oxygen as a source of energy and therefore make energy through fermentation. Depending on the type of fermentation taking place, homo- and heterofermentative *Lactobacillus* spp. can be found.

The temperature at which *Lb. plantarum* and *Pediococcus* spp. ferment sugar most efficiently is around 35°C, but this temperature is generally not found during fermentation except in countries such as the United States, where temperature between 35 and 38°C are frequently applied. *Lb. sake* and *Lb. curvatus* work at lower temperatures, around 20–25°C. *Lactobacillus* spp. generally do not produce the enzyme catalase even though most of them produce H_2O_2. *Lb. sake*, *Lb. viridescens*, and *Lb. curvatus* are known producers of H_2O_2, a substance that is not desirable to produce during fermentation of salami. Different species of *Lactobacillus* added to salami produce sufficient amounts of lactic acid to acidify the salami adequately in a fairly short period of time. They are also able to grow at low temperatures of between 12 and 28°C. Fermentation of sugar by certain *Lactobacillus* spp. at 20–25°C results in sufficient levels of lactic acid and the acid taste obtained is very acceptable. *Pediococcus* spp. require temperatures between 30 and 35°C and a strain of *Pedicoccus acidilactici* is used for many years in applications when temperatures of around 38°C are applied. High-temperature fermenting cultures are commonly applied in the United States for technological and historical reasons in products such as summer sausage and pepperoni salami.

Most LAB produce bacteriocins, proteins, or peptides, which are effective against some pathogenic microorganisms. Bacteriocins are not antibiotics but are digested, or broken down, to amino acids in the digestive system. One strain of *Lb. sake* produces a bacteriocin called sacaicin, which is effective against *Listeria monocytogenes*. *Lactococcus lactis* produces nisin, which is effective against *Staphylococcus aureus* as well as *L. monocytogenes*.

Other LAB widely used within the production of salami are members of the genus *Pediococcus*. The most common species are *Pediococcus (P) acidilactici*, *P. pentosaceus*, and *P. cerevisiae*. *P. acidilactici* ferments sugars fastest at temperatures around 40°C but such a high temperature is rarely used during production of salami, except in the production of summer sausage in the United States. *P. acidilactici* forms lactic acid from glucose, galactose, arabinose, and xylose. *Pediococcus* spp. are generally added at levels of 10^5–10^6/g of salami. *Pediococcus* spp. introduced into the salami mass die soon after the product acidifies, whereas *Lactobacillus* remain alive. *Pediococcus* spp. contribute more significantly to a better overall salami flavor than *Lactobacillus* spp. as *Pedicoccus* spp. have some proteolytic activity. *P. cerevisiae* grows optimally at 25°C and its thermal death point is 60°C. No gas is formed by *P. cerevisiae* when fermenting dextrose and maltose.

Members of the family Micrococcaceae such as *Staphylococcus (St.) carnosus*, *St. xylosus*, and *Micrococcus (M.) varians* (now known as *Kokuria varians*), *M. candidus*, and *M. aquatilis* are added as starter cultures as well. Most of the *Staphylococcus* spp. and *Micrococcus* spp. added do not produce acid at all. Almost all *Micrococcus* spp. are strictly aerobic bacteria and are only weakly active during fermentation. Given that they are aerobic bacteria, they cannot act effectively in the core of a salami since there is insufficient oxygen available, or none at all. *Staphylococcus* spp. are aerobic as well as anaerobic and are active in the core of a salami. Because of this, *Microsoccus* spp. are most often added in combination with *Staphylococcus* spp. Adding *Micrococcus* spp. can be advantageous as their nitrate reducing activity does not depend as much on temperature as that of *Staphylococcus* spp. does.

Micrococcus spp. also produce the enzyme catalase and being catalase positive supports the curing color as well as the aroma of the product and protects the product against oxygen. *Micrococcaceae* are added at 10^6–10^7/g of sausage. The enzyme catalase breaks down the viscous H_2O_2 into water and oxygen and therefore neutralizes this highly reactive material.

The presence of H_2O_2 can result in discoloration of the salami given that H_2O_2 is a very strong oxidizing agent and can bleach myoglobin resulting in a green-yellow color of the salami. H_2O_2 also leads to the formation of radicals and therefore speeds up rancidity. By catalase breaking down H_2O_2 rancidity is delayed as well.

Micrococcus spp. and *Staphylococcus* spp. generally produce the enzyme nitrate reductase, which is of great importance in obtaining and stabilizing a strong curing color. Large amounts of added nitrite are oxidized to nitrate during the manufacturing process and nitrate would not contribute to color unless it were reduced to nitrite again by the enzyme nitrate reductase. *Micrococcus* spp. generally do not cause much lactic acid to form and therefore they should not be applied on their own for purposes of acidification. Often, *Micrococcus* spp. is applied in combination with acid-producing starter cultures. By using this combination, the product acidifies as required and the *Micrococcus* spp. reduces nitrate to nitrite for development of flavor within the product.

Staphylococcus spp. are facultatively anaerobic and their metabolism is either respiratory or fermentative. Under anaerobic conditions, they produce lactic acid from glucose, while under aerobic conditions, the main products they produce are acetic acid and CO_2. *Staphylococcus* spp. grow better under aerobic conditions. *Staphylococcus xylosus* shows a wider range of enzyme activity than *St. carnosus* and contributes significantly to the development of a good flavor in a salami. Members of both the *Staphylococcus* and *Micrococcus* families cause lipolysis as well as proteolysis to some degree. Lipolysis (breakdown of fat into free fatty acids) and proteolysis (breakdown of proteins into free amino acids) occur during fermentation and drying and are largely responsible for the development of the typical salami flavor. A combination of *Lactobacillus* spp. and *Micrococcus* spp. is frequently added in salami given that addition of *Lactobacillus* spp. on their own makes for shortfalls in color, flavor, and aroma in the finished product. If *Micrococcus* spp. are added, they benefit color and flavor development in fast and medium-fast fermented salami.

Countless different blends of starter cultures are on the market today containing a mix of bacteria, and these are very often used instead of adding a single species of microorganism. The mixture is chosen based on the fermentation process desired. A distinction is made between starter cultures for fast, medium-fast, and slow acidification. Fast starter cultures are used to produce salami, when the pH has to drop to 5.2 (and below) within around 24–48 hours. Medium-fast starter cultures are used when the pH needs to drop below 5.2 after around 48–96 hours. The starter cultures used for production of slow-fermented salami are most often added as protecting cultures and do not contribute to acidification at all. Their purpose rather is to further development of curing color and flavor within the product. Generally, a fast drop in pH means that microbiological stability is achieved quickly, but the flavor and taste of the product commonly suffer.

7.2.3 Selected Wanted Mold and Yeast

Yeast and mold are aerobic fungi and can only exist on the surface of salami. Both yeast and mold (noble mold) are introduced to the surface of salami via inoculation. They are either sprayed onto the product or the product is dipped into a solution containing the mold or yeast.

In terms of the molds applied, members of the genus *Penicilium* are frequently chosen, for example, *Penicilium nalgiovense*. In terms of the yeasts used, members of the genus *Debaryomyces* are occasionally applied, and *Debaryomyces hansenii* is commonly the preferred choice. *Candita famata* is another yeast added to salami. Members of these families are chosen as they are salt tolerant but do not reduce nitrate to nitrite.

Yeast requires oxygen for growth and only grows on the surface of a salami. Even slight treatment with smoke kills yeast present on the surface as yeasts cannot tolerate certain components in smoke, such as phenols and organic acids. Yeast and mold protect salami against the influence of oxygen and stabilize the color. Exposing the surface to less oxygen and light also slows down development of rancidity. Last, formation of a dry ring, or case hardening, is avoided as the layer of yeast or mold protects the surface from excessive drying out.

Some strains of *Debaryomyces* are very tolerant against a low Aw and even grow at an Aw of 0.86. There is evidence that *Debaryomyces hansenii* inhibits growth of *St. aureus*. *Debaryomyces hansenii* helps the development of a stable and strong curing color as well as the development of the typical salami flavor inside the salami and is added directly in to the sausage mass. *D. hansenii* neutralizes lactic acid and contributes to a milder flavor. It also supports the breakdown of protein into peptides and amino acids through proteolysis and the formation of free fatty acids through lipolysis, contributing to formation of a good flavor.

Molds such as *Penicilium (P)* cause a white or gray-white mycelium to grow on the surface of salami. Their conidia should not be green, black, or yellow. Mold present on the surface of salami keeps away oxygen and therefore stabilizes curing color inside the salami. Mold also contributes to the development of the typical salami flavor as it contains proteases and lipases, which break down proteins and fats into components such as amino and fatty acids. In addition, the presence of the layer of mold on the surface slows down development of rancidity as it protects the salami from the impact of oxygen, as mentioned, and light. The layer of mold also slows down the loss in weight during drying of the product and minimizes the risk of obtaining a dry ring (case hardening).

Molds commonly added are *Penicilium nalgiovense* and *P. candidum*. Neither of these produce any mycotoxins and all noble molds applied to salami must be nontoxic. The color of the conidia and mycelia growing on the surface of the product must be white or gray-white. Noble mold must grow fast, cover the sausage uniformly (within 3–5 days), and develop a specific and consistent flavor. The added mold also must not contain any enzymes active on cellulose as these would literally "eat" up fibrous casings. Nonwaterproof materials such as fibrous casings are used for products with surface mold. Most of the casings chosen have a rough surface for the mold to cling onto, which makes them grow well. Occasionally, the fully dried salami is rolled in talcum powder at the end of the drying process to compensate for uneven growth of mold during drying.

7.2.3.1 Prevention of Unwanted Mold

Unwanted mold can cause a series of problems in the production of salami. While noble mold is white, or gray-white in color, unwanted mold can display a wide range of colors and is often black, green, or even yellowish. The major cause for growth of unwanted mold is having a high level of RH for too long of a period of time at the beginning stage of fermentation and/or the application of smoke too late in the production process. If these conditions occur, generally, mold starts to grow on salami after around 3–4 days during the initial stage of fermentation. Air flow has an impact on mold growth as well and if air flow is too slow, mold growth increases. Growth of mold also occurs during drying of products, once again usually because the RH is too high and the air velocity is too slow. Drying rooms should be free of mold given that mold here would inoculate mold-free products and spread the problem around.

Besides being not attractive to the human eye and producing a bad smell, a few species of unwanted mold produce different mycotoxins, which penetrate the product. Therefore, the removal of unwanted mold from the surface of the salami only removes the visible part of mold: the mycotoxin is still present inside the salami. Proteolytic mold can cause a sausage casing to deteriorate entirely. Black spots can be occasionally seen on the surface of a salami just underneath the casing as a result of the presence of unwanted mold. This is possible as those areas within the sausage mass just underneath the casing experience aerobic circumstances, which allow mold to develop.

The two common ways of preventing growth of unwanted mold is to apply natamycin or potassium sorbate to the salami. However, it is best to avoid growth of unwanted mold during the initial stages of fermentation in the first place, mainly by controlling the level of RH during all stages of fermentation and drying and applying smoke at the correct time. Growth of unwanted mold can be fought successfully using natamycin, an antifungal produced by *Streptomyces natalensis*. Natamycin forms a complex with the steroids found in molds and yeasts and destroys their cell membrane. As a result, permeable cell membranes permit cations and other ions to penetrate into the mold cell, making its pH drop rapidly and ultimately, the cell dies. A major advantage of natamycin, compared with potassium sorbate, is that it does not penetrate the sausage mass at all and therefore has no impact on fermentation, color, or flavor development, especially in the outer layers of the product. Steroids are not present in bacteria, which means that natamycin has no impact whatsoever on the bacteria present in salami. It has no impact on the starter cultures added and the entire fermentation process is not influenced in any way. There are a few products on the market containing natamycin. In most cases 5–7 g of a commercial product is mixed with 1 L of warm water, well stirred, and then left standing for around 30 minutes. During this time the mixture will thicken slightly. Around 8–10% of salt is added afterward and this final mixture can be sprayed onto the salami or the freshly filled salami can be dipped once into this low-viscosity slurry. Experience shows that natamycin is several times more effective than potassium sorbate.

Another option to avoid growth of unwanted mold is to apply potassium sorbate. The filled salami can be either dipped right after filling in a 3–4% potassium sorbate solution or the filled salami can be sprayed with this solution 1–2 days after fermentations has commenced. Most commonly, the filled salami is dipped into the solution for a few seconds and spraying is not commonly practiced. Several food standards in the world have a set maximum level of residual sorbate within the outer layers of raw fermented salami and the strength and duration of the dipping solution as well as the actual dipping process has to be adjusted so maximum levels in the finished product. The disadvantage of using sorbate is that it penetrates into the outer layers of the salami and can interfere with fermentation and cause discoloration. Another method applied to reduce growth of unwanted mold is to add 1–2% of potassium sorbate into the soaking solution, which casings are placed into before being filled (see Section 7.3.4).

Mold normally grows at RH levels above 75% by warm temperatures and low air velocities speeding up growth. High levels of humidity favor the growth of black molds and proteolytic molds, which "eat up" casings. If unwanted mold is present in fermentation or drying rooms, they should be cleaned with an alkaline cleaning detergent as acid-based cleaning materials are not as effective as alkaline materials against molds. Once the room is properly cleaned and rinsed, a solution containing 5–7% of potassium sorbate can be sprayed all around it. The sprayed room is then left standing to dry completely and the sorbate solution is not washed off. Finally, all the tools used during cleaning should be disposed given they would cause regrowth of mold if they were used again. Salami affected by growth of mold can also be washed with a salt or lactic acid solution and cold smoked once it is dry again to remove unwanted mold (if the product is not smoked afterward, then the mold grows again and is only removed for a couple of days). However, as said earlier growth of unwanted mold should be avoided in first place as this makes labor-intensive steps such as washing unnecessary. Vacuum packing of the finished product or an RH below 75% during drying suppress growth of mold during storage.

7.3 MANUFACTURING TECHNOLOGY

During manufacturing the carefully selected raw materials are combined with the chosen additives to create a safe, tasty, and strong-colored end product with good slice coherency. Modern technology facilitates the production of a perfect salami, but "additives" such as experience and sound knowledge of the cutting, mixing, fermentation, and drying processes are still of great help.

7.3.1 Salami Made in a Bowl-Cutter

Salami with a particle size of 1–3 mm is commonly produced in a bowl-cutter giving the end product the typical appearance because not all particles are exactly of the same size is it is the case if a mincer-filler head system (see Section 7.3.3) is used. The bowl-cutter must not be rusty as any rust introduced into the salami mass interferes badly with fermentation and color development. The bowl-cutter must also have been cleaned properly the previous day as rapid bacterial growth would have taken place in any particles of sausage remaining in the bowl-cutter and all those bacteria would subsequently be introduced into the new batch of salami. In addition, no traces of cleaning and/or disinfection materials must be present on the machine as they can interfere badly with fermentation.

The knives of the bowl-cutter should be sharp given that a clear cut of the meat and fat materials is desired and smearing of fat and meat during the cutting process is to be avoided: smearing of fat causes major problems during fermentation and drying of the product. The sharpening angle of knives used for cutting salami should be around 22–25 degrees (a small angle produces a sharp knife). The knives also have to be adjusted so that the distance between the end of the knives and the bowl of the bowl-cutter is a maximum of 1–2 mm. The number of knives present varies between three to six but six knives is the norm. Bowl-cutter knives rotate at around 55–70 m/s or 1200–1500 rev/min (rpm) during cutting and are slowed down to around 3–5 m/s during mixing. Special salami bowl-cutters are on the market, which have two sets of knives. These double-cutters have the advantage that the salami mass is cut twice during one turn of the bowl and the desired granulation is reached quickly. Obtaining the desired particle size in a short period of time lowers the risk of smearing of fat once again and more batches of product can be produced within a certain period time, which is economically beneficial.

In general, the bowl-cutter should be around half-full during cutting of the meat and fat materials to ensure a proper flow of material to be cut. More specifically, if a 200-L bowl-cutter is used for production, the batch size should be between 100 and 120 kg. Overloading of the cutter leads to jamming and the materials do not get cut cleanly. Jamming also increases the temperature of the mass rapidly resulting in an enhanced risk of fat smearing. As mentioned, fat and meat materials are generally cut under medium-fast knife speed at around 1200–1500 rpm, or 55–70 m/s (slower than during the production of an emulsion by cooked sausages) and the bowl speed is around 8–12 rpm to ensure that the materials to be cut flow "freely" through the cutter. During mixing of the mass in the final stage, the knife speed is reduced drastically to around 2–4 m/s or around 200–300 rpm.

Different methods of cutting a salami in the bowl-cutter are commonly practiced:

1. Hard frozen fat, and here primarily pork back fat, is cut into smaller chunks with the frozen meat cutter and placed in the bowl-cutter, and the fat is cut for a while until a granulation of around 8–10 mm is obtained. The reason for cutting fat first for a while on its own is that frozen fat seems to "move away" from the fast-turning knives and if fat and meat are added at the same time, the meat is more finely comminuted than the fat. Semifrozen or tempered meat showing a temperature between −8 and −5°C is added afterward into the cutter and additives such as starter cultures, spices, and color enhancer are added evenly while cutting under medium-fast speed. Hard frozen meat is not used as it would reduce the sharpness of the knives very quickly given that lean meat contains around 75% and cutting frozen meat can be compared with cutting of ice. If this type of meat were cut, they would need to be resharpened very often. Predried meat, as described under Section 7.1.1, demonstrating a temperature around −2°C is commonly added as well.

All additives, spices, as well as starter cultures have to added evenly during cutting as a batch of salami experiences significantly less cutting and mixing than, for example, a batch of finely cut emulsified sausage. Unevenly introduced additives, especially starter cultures, are commonly the cause of faulty salami and a lack of fermentation can be the result of such inaccuracy. During this cutting process the temperature of the meat mass in the bowl-cutter should be between −4 and −6°C. Once the desired granulation, or particle size, is obtained the knife speed is reduced to around 200–300 rpm and salt as well as nitrite are evenly introduced to guarantee even and secure color development. Besides salt and nitrite around 15–20% of chilled minced (3 mm plate) meat is also added to the bowl-cutter to raise the temperature of the sausage mass.

The salami mass is just gently mixed afterward until the salami mass is slightly tacky and a small degree of protein is activated by the presence of salt. This tiny amount of activated protein is called a "sol," a substance made from salt, activated protein, and water (from the meat itself). The sol is needed for slice coherency during fermentation as well as subsequent drying and it is a colloidal system where particles within are still moving freely. (A gel, in contrast, is a solid substance in, which particles do not move freely.) The finished salami mass should be at a temperature of between −3 and −1°C and a combination of fully frozen fat, tempered and predried meat, as well as some chilled minced meat is selected so that the salami mass ends up in this temperature range. When no predried meat is available, then frozen meat is tempered so that a combination of frozen fat and tempered meat results in the same temperature within the finished salami mass.

2. Another method of cutting salami is to combine frozen fat, semifrozen fatty meat material, as well as some chilled meat in the bowl-cutter at the same time and to commence cutting under medium-fast speed. All additives including starter cultures except salt and nitrite are added evenly into the sausage mass right at the beginning of the cutting process and all the sausage mass is cut until the desired granulation is reached. At this point, salt and nitrite are added evenly, and the materials are mixed for a short time until some sol is obtained and the mixture is slightly tacky. Experienced producers of salami have worked out the ratio of frozen semifrozen and chilled material needed to obtain a final temperature of around −4 to −1°C in the salami mass. Generally, the addition of salt at the end of the cutting process reduces the temperature of the entire sausage mass by around 2°C.

Even though the temperature is as low as −4°C, water in meat does not freeze again due to the fairly high amount of added salt to the sausage mass (25–30 g/kg). The added salt lowers the freezing point to around −4°C. Temperatures below −4°C would cause the water to turn into ice and no sol would form, resulting in poor slice coherency. Pores would also be present in the finished product as full vacuum cannot be obtained during the filling process.

During all these different cutting methods, the coldness drawn from semifrozen meat materials supports the "clean" cutting of fat: the surface of the cut fat remains frozen due to the presence of the semifrozen meat and cleanly cut meat and fat particles can be seen as a result in the end product.

Pure beef salami is commonly produced from fatty beef brisket and the meat material is used in a semifrozen state and cut until a particle size of around 2–3 mm is obtained. All additives except salt and nitrite are added during the beginning stage of the cutting process. Salt and nitrite are added at the end of the process and a final temperature of the sausage mass between −4 and −1°C is desired.

As a general rule, the finished mass of a finely cut salami should be slightly tacky once removed from the cutter. The temperature of the mass should be between −4 and −1°C once the cutting process is completed to eliminate the risk of fat smearing during cutting. Salami with a smaller particle size should have a temperature of around −3 to −4°C while salamis with larger particle sizes can have temperatures of −3 to −1°C. In any case, the temperature of the sausage mass should be at or preferably below 0°C. Super-fine cut salami, with a particle size of around 0.8–1.0 mm, is cut using dry-ice (CO_2). The dry ice is gradually added into the bowl-cutter during cutting to keep the temperature of the mass well below 0°C during the prolonged period of cutting. This is so that small particles are produced without fat smearing. The CO_2 maintains the temperature at around −4 to −2°C during cutting. Once the desired granulation is reached, salt and nitrite are evenly mixed into the sausage mass until some degree of binding takes place (ie, a sol forms).

CO_2 is 1.5 times heavier than air and it is a linear molecule. It condenses under normal pressure directly into a solid at around −78°C and this solid is known as dry ice. Solid CO_2 does not melt under normal atmospheric pressure but sublimes directly into a gas (sublimation is a process where a substance turns from being solid directly into a gas without having been present as a liquid). Liquid nitrogen can be used for cooling purposes as well but the introduction of nitrogen into the salami mass has to be tightly controlled as nitrogen has an enormous cooling capacity and incorrect introduction of nitrogen into the salami mass can reduce the overall temperature of the sausage mass easily to levels of −10 to −15°C. Nitrogen (N_2) is made from liquefied air and has a boiling temp of −198°C. During evaporation, large amounts of heat are removed and N_2 is an inert gas, meaning it does not react with any component within the sausage mass.

Smearing of fat has to be avoided during cutting as this would block capillaries in the sausage mass and subsequent drying would be badly affected. The period of time between obtaining the finished mass in the bowl-cutter and filing of the salami should be a short as possible. If a batch of salamis is unloaded into a trolley and left standing for an extended period of time, the internal temperature will drop even further and can go below −5°C. If that is the case, the water present turns into ice. Pores will therefore be seen in the final product and the filling process will very difficult to achieve. The semifrozen sausage mass has to be mixed again in the cutter to raise the temperature to remove possible lumps of frozen material within the sausage mass. Generally, the person working the bowl-cutter should work at the speed at, which the products are being filled to avoid the sausage mass resting for a long time before filling. If the sausage mass is frozen on filling, the vacuum applied does not remove entrapped air properly and therefore there will be pores in the final product. The application of a slight vacuum during the final mixing stage of the cut sausage mass is advantageous as more sol is obtained, which benefits sliceability in the final product. As oxygen is removed, development of the curing color is enhanced and the Eh value is reduced within the sausage mass as well. Reducing the Eh value provides aerobic spoilage bacteria such as *Pseudomonas* spp. with less oxygen thus inhibiting their growth.

7.3.2 Salami Made With Mincer-Mixing System

Salamis with a particle size of 4–13 mm (or larger) are generally produced by mincing the meat and fat materials and then mixing them properly. Pure beef salami, predominantly made from fatty beef briskets, can be produced in this way as well. If a bowl-cutter is available, all the meat and fat materials are placed in the cutter and cut up only to for a few rounds to obtain pieces of meat and fat, which can subsequently be minced. During the few rounds of cutting (which is done with a slow knife speed), starter cultures (if used) are evenly mixed into the sausage mass. The temperature of the sausage mass at this point is around −2 to 0°C. A combination of frozen, semifrozen, preconditioned, and occasionally chilled materials are used to obtain the desired temperature overall. After removal from the cutter, the mass is minced with a plate, which produces meat and fat particles of a diameter between 4 and 8 mm. It is vital that mincing is carried out without any fat smearing taking place and that cleanly cut particles of meat and fat are obtained.

The minced materials then dropped onto a belt. All other additives such as spices, color enhancer, nitrite, and salt are added by being dropped onto the minced materials from a dispenser placed above the belt as the materials pass by. The belt leads to a mixing device and the minced materials and the additives on top of them drop into this mixing device, which has several small fast rotating arms. The whirling action of the rotating arms distributes all the additives well among the minced meat and fat material. The material is then placed into a paddle-mixer for final mixing, Final mixing takes place gently for a short period of time to distribute all additives evenly and to obtain a sol to some degree. It is also helpful to apply a vacuum during the final mixing stage to remove trapped air. This makes color development take place faster and the Eh value is reduced as well.

Large processors of minced salami operate fully automized lines in which frozen and semifrozen materials are cut with a mincer into particles of around 13- to 20-mm diameter and then dropped into a large paddle-mixing machine. All additives including starter cultures are added and the mass is mixed for a short while. The mixed materials drop onto another belt and are then transported to the mincer where final mincing takes place. The minced materials drop onto another belt, which loads a paddle-mixer within which final mixing takes place. In some cases, the mass drops from the belt onto a whirl-mixer before landing inside the final mixer.

Another method of producing minced salami is to place all the materials in the bowl-cutter, put in all the additives and starter cultures but not the salt and nitrite, and then cut the materials under slow knife speed until minceable pieces of meat and fat are obtained. The coarse sausage mass is removed from the cutter and minced with the desired blade. Following mincing (which should be carried out without causing smearing), the mass is mixed in a paddle-mixer by adding salt and nitrite until some degree of tackiness is seen. Paddle-mixers are used as they do not cut the individual particles of meat and fat any further and produce an efficient gentle mixing action. As under Section 7.3.1, a long resting time between mixing of the salami mass and filling should be avoided.

7.3.3 Salami Made With Mincer-Filling Head

This method of production is commonly applied to keep the process of producing salami as simple as possible. All meat and fat materials are minced with a plate, which is generally twice as large as the final particle size in the end product. Specifically, if the desired final particle size is 4 mm in the salami meat and fat is minced with the 8–10 mm plate in the first place. Fat should be hard frozen when processed while other meat and fatty trimmings can be tempered and chilled. The temperature of the minced meat mass should be between −2 and 0°C and the minced materials are then loaded into a paddle-mixer. All additives, cultures (if applied), spices and salt are added and all is mixed for a short while to ensure even distribution of all materials as well as obtaining some degree of tackiness. The mixed meat mass is subsequently loaded on to a vacuum filling machine, which has a mincing device attached to the filling machine. During the filling process a vacuum is created first before the meat mass is minced to its final particle size. The minced meat and fat materials are finally filled into the respective casing and the mincing and filling process takes place as one continues process. The advantages of such a process are that the particles of meat and fat are consistently of the same size and this process does not require highly skilled staff. Care has to be taken to ensure proper mixing before mincing/filling as no further mixing takes place during the mincing/filling process. An unevenly mixed meat mass will display faults such as an uneven curing color or even nonfermented areas within products. Modern processing technology nowadays allows salami to be produced following this process with a particle size as small as 0.8 mm without smearing of fat. Obtaining the correct temperature of the mixed meat mass after mixing and before filling is of vital importance. If the temperature is too warm smearing of fat will be obtained during mincing and filling affecting drying negatively. In case the temperature is too cold the mincing-head can jam or block-up. As a general rule, the temperature of the mixed meat mass before filling should be colder if the particle size in the final product is to be small. On the other hand, if the final particle size is, for example, 6–8 mm, temperatures between 0°C and −2°C can be tolerated.

In case no mincer is available for the first mincing step a bowl-cutter can be used instead to cut all meat and fat materials to the desired particle size followed by the addition of all additives and spices. All is then mixed for a short while until some degree of tackiness is obtained before the mixed salami mass is passed through the mincer-filling head. If possible a bowl-cutter should be used, which has the option of a mixing-speed as well as mixing with reverse knife speed; that is, the knives on the bowl-cutter turn backward, ensuring proper mixing of the meat mass without further cutting. Within this application, the mincer is replaced with a bowl-cutter.

7.3.4 Filling

Filling should take place right after the sausage mass is produced as long resting times lead to a decrease in temperature within the salami mass and the appearance of hard lumps due to the formation of ice. The cut or minced/mixed salami mass is filled into permeable casings. Fibrous and collagen casings are most commonly chosen. Salamis filled into natural casings have a very attractive Old World appearance, but care must be taken that the natural casings have a low bacteria count and that all adherent fat has been removed as the salami would otherwise become rancid quickly. Natural casings, besides their attractive appearance, protect the sausage more effectively against case hardening compared with fibrous and other types of casing given that they contain a fair amount of fat and connective tissue, which acts a moisture buffer.

All casings used, regardless of origin, should be treated in the correct way as per manufacturer's recommendations to ensure maximum functionality of the casing. It is important that the casing is soaked (if so) for the recommended period of time and filled to the desired diameter, shrinks with the product during drying, and clings to the product in the desired way. Some fibrous casings used today contain antimold substances such as food acids and sorbate, which reduce the risk of unwanted mold forming during fermentation and subsequent drying. Fibrous casings should be soaked for 20–30 minutes in water exhibiting a temperature between 30 and 40°C in order to secure proper stretch and shrinkability of the casing. This is especially important if shirred casings are processed because shirring compresses the casing tube and sufficient soaking time has to be provided for the water to reach all areas of the compressed casing. Care has also to be taken when insufficient casings have been soaked in the first place and additional casings are placed in the soaking water together with casings, which have been soaked for longer time already. If such newly added, or undersoaked, casings are filled, the casing will not demonstrate the desired stretch and shrinking behavior. Salami should never be filled into waterproof casings as no smoke would penetrate through the casing onto the product. The product would also not dry, the process during which moisture is removed from the salami. Some salami are filled into molds of triangular or other shapes. These molds have to be lined with a breathable foil first to avoid the sausage mass sticking to the mold. Without the foil, subsequent removal of the sausage from the cage, or mold, would be impossible.

The sausage mass at point of filling should have a temperature between −4 and 0°C depending on the particle size to avoid fat smearing during filling. Fat smearing creates a thin layer of fat underneath the casing and drying is badly affected as capillaries are closed by the layer of fat. In extreme cases, drying can be inhibited altogether by fat smearing. A high Aw for long period of time in the core of the salami can lead to microbiological spoilage as unwanted bacteria such as *Salmonella* spp. have prolonged access to free water until the Aw drops to 0.95, the Aw by which growth is inhibited. If the sausage mass is at temperatures below −4°C, it is extremely hard to handle for the filling machines. A proper vacuum also cannot be applied during the filling process at temperatures below −4°C due to ice formation within the salami. Therefore, pores, or small air pockets, can be seen in the final product, which is not desired. Another reason for applying full vacuum during filling is because air introduced into a cold salami mass during cutting or mixing requires string vacuum action to remove entrapped from a firm meat mass.

The filling speed should be moderate given that fast filling speeds increase friction within the sausage mass while it passes through the filling horn (stuffing pipe). This increases the risk of fat smearing. The largest possible diameter filling pipe in relation to the casing should be used as a large diameter lowers friction in the sausage mass as the mass passes through the pipe and to avoid back-curling of the sausage mass. The filling pipe should also be as short as possible to reduce the degree to which the sausage mass is squeezed as it passes through the filling horn.

Generally, salami is filled tightly into casings and creating a vacuum during filling is of great benefit. Due to the absence of oxygen, a better curing color is obtained and the reduced Eh value gives unwanted aerobic bacteria spoilage bacteria less chance to grow. A more compact sausage is obtained if a vacuum is created during filling and no pores are therefore visible in the finished product. Large amounts of oxygen are introduced into a salami mass while it is mixed in the bowl-cutter or mixing machine and high levels of oxymyoglobin form. In fast fermented salami, the oxymyoglobin (especially in small-diameter products) are liable to denature and this results in an atypical color in the finished product. If oxygen is removed during filling, this aids the formation of nitrosomyoglobin, which helps to avoid this problem. Modern filling machines redirect the sausage mass very little during the actual filling process, which is a great advantage. Older-style filling machines usually redirect the sausage mass and this increases friction and therefore the risk of fat smearing in the sausage mass is greater. Occasionally, the salami mass is filled first into cylindrical barrels and then the filling horn is attached to the front of the barrel. The sausage mass is pushed from the opening at the other end of the barrel toward the filling horn. This means that the sausage mass is not redirected during filling, which reduces the risk of fat-smearing. This process takes place under a vacuum.

Some filling machines are combined with a mincing device and so mincing and filling take place at the same time (see Section 7.3.3). The surfaces of working-tables and benches used during filling must be clean as any contamination would lead to defects in the product to be fermented such as poor curing color, discoloration, microbiological spoilage, and growth of mold. Sausage mass produced with GDL must not be left for a long time before filling as acidification starts as soon as GDL comes in contact with water from meat. The transformation of GDL into gluconic acid is a chemical reaction and the formation of gluconic acid causes the pH to drop. Once the pH value of the sausage mass drops below 5.2 (which mainly happens on the surface of the sausage mass placed in trollies as the surface is warmer than the core), the sol is transformed into a gel, which is then destroyed during filling. Once a gel is destroyed, it will not form again and poor slice coherency will be the result. Under no circumstances must an unfilled salami mass, produced with GDL or fast-acting starter cultures, be placed in the chiller overnight or be left standing anywhere else. It this occurs, the pH value within the entire sausage mass could drop below 5.2 and all sol would be transformed into gel before the product finally was filled into its casing.

Cupping is a commonly observed problem in small-diameter pizza salami. This is when the outer areas of a slice of salami move away from the pizza under the severe impact of heat in the pizza oven, but the center remains attached to the pizza, forming a cup shape. This problem can largely be prevented by using the largest filling pipe possible for the respective casing and by filling the salami mass straight into the casing, avoiding back-rolling (back-curling). To be more specific, the salami mass should not be redirected in any way as it is filled into the casing. When a small-diameter filling horn is used to fill a casing of a much larger diameter, the sausage mass curls back and is mixed up again, which leads to cupping on the pizza. When filing all types of salami, the largest filling pipe possible for the chosen casing diameter should be used (Fig. 7.1).

Clips have to be attached firmly to the casing covering all the casing material avoiding the salami to fall of the stick. If clips are attached too firmly it can "cut" the casing, causing the salami to drop to the floor. On the other hand, if the clip it attached loosely, the pressure coming from the tightly filled salami will cause the clip to move outward and the clip is lost totally. In such a case, the salami loses its tension and a visually unattractive end product is obtained. It is important that the correct clip is chosen based on the thickness of the casing because, as said earlier, the clip has to cover the bundle of casing properly to "close" the casing fully, also ensuring that the clip does not move on the casing itself, maintaining tension on the filled product.

7.3.5 Fermentation

There are two significantly different systems in place in regard to fermentation and drying. One system could be called the "Italian" system, while the other one is known as the "German" system. The Italian system works on the principle that the moisture from the product in the room generates the moisture required for proper fermentation and drying. It also functions in a way that processes such as dehumidifying of the room or air velocity inside the room are not constant processes. More specifically, when products are placed into the room, a minimum and maximum value is set in regard to temperature and RH. Air velocity is set as well and the room is heated to the set temperature. Once

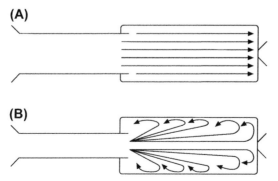

FIGURE 7.1 Using a wide filling horn (A). Using a small-diameter filling horn (B) results in the salami mass being redirected (*curling back*).

the preset temperature inside the room is reached, the room "shuts off" and the moisture from the product causes the moisture level in the room to increase. Dehumidification in the room only starts as soon as the moisture level reaches the set maximum level and continues until the set minimum level is obtained. The set minimum level of moisture is less than the moisture level in the product. Then, once again, the room shuts off, letting the moisture from the product to be released into the air to increase the level of moisture in the room. This "Italian" system is most often applied by salami that should exhibit noble mold on the surface because periods of no air movement inside the room support growth of mold. The Italian system, because of the resting periods during humidification, is also gentler on the product during drying and is preferred for high-quality salami products. This Italian system works well if the room is full with products because large volumes of product increase humidity relatively fast inside the room, resulting in short resting periods before dehumidification needs to restart. However, if little product is inside the room, it can take a long time until the set maximum value toward humidity is reached before dehumidification starts again. In such a case, the minimum and maximum set points toward the RH inside the room are kept not far apart, causing the room to dehumidify more frequently, resulting in efficient drying of the product.

The "German" system is based on the principle that there is always some degree of air flow within the fermentation or drying room and as such does not allow for any resting periods with no air flow present inside the room. This system is preferred when mold is not desired to be present on the end product as well as if the desired weight loss should take place in the shortest possible time.

Fermentation is the process in which the raw and microbiologically unstable sausage mass turns into a shelf stable product with strong curing color, good sliceability, pleasant flavor, and, most important, microbiological stability. A raw fermented salami can be made microbiologically stable in two ways: firly by having an Aw of or below 0.89 (via drying) and, second, by having a pH value below 5.2 (via acidification using GDL, citric acid, or starter cultures, which ferment sugars into lactic acid). In some salami produced, both hurdles are in place at the same time, which makes the product particularly safe. One hurdle, though, is sufficient to control bacteria such as *Salmonella* spp. and *S. aureus*, which are the greatest hazards in salami. Food standards in some countries demand a decline in pH value below 5.2 within 48 hours after fermentation commenced to reduce the risk of growth of *Salmonalla* spp. and other bacteria such as *L. monocytogenes*.

Products made from pork and beef, or only pork, ferment faster than products containing only beef. This is because pork usually contains more lactic acid than beef and beef generally has a higher initial pH value than pork as well. Beef also generally has a larger buffer capacity, which reduces total acidification during fermentation. The level of fat present within the product also plays a role in fermentation. Salami with a lower fat content automatically contains elevated levels of lean meat. An increased level of meat within the product raises the level of water and therefore the Aw. As a result, more water is available to bacteria such as *Lactobacillus* spp., fermentation is prolonged and the pH value drops slightly lower (and faster) than it would do in a fattier product, with less free water. The underlying theory here is the activity of most LAB comes to a halt at an Aw around 0.95. If the Aw of the salami mass is above that figure for a longer period of time, bacteria that cause acidification will be active for longer. In addition, it is usually said that increasing the amount of lean meat in the sausage mass results in higher levels of glycogen and this also leads to a greater degree of acidification. It may be the case, though, that increasing the level of lean meat also increases the buffer capacity of proteins and more sugar, or GDL, has to be added for the desired drop in pH to take place. On the other hand, elevated levels of fat reduce the Aw of the freshly produced sausage mass and the RH has to be higher during fermentation, especially within the first 24–48 hours, to avoid case hardening.

During fermentation and drying, the conditions in the smoking chamber or drying room such as temperature, air velocity (speed of air flow), and RH greatly determine the reactions inside a salami. The conditions in the salami, in turn, determine factors such as color and color development, texture, weight loss, flavor, and microbiological stability of the product. The fermentation room should be free of mold and clean before it is loaded with freshly filled salami. Salami generally have an Aw of 0.96–0.97 when they are filled. "Greening" and "maturing" are other terms used interchangeably with fermentation, although greening more specifically can describe the first 48–76 hours of fermentation until the pH value has dropped to the desired level. After fermentation, drying or maturing, is the next and final step.

Fermentation should only commence once the fermentation room has been loaded to ensure that all freshly filled products are exposed to the same climatic conditions right from the start. The filling, or loading, of a fermentation room (depending on its size and how fast trollies are placed into it) can last for hours. There should also be no air flow during loading of the fermentation room as prolonged periods with air flow at an RH around 75% can create case hardening before fermentation has even started. Only once the fermentation room is full, or production of salami has finished and all the product has been loaded into the fermentation room, can the fermentation program begin.

The filling level of the room also has a significant impact on fermentation and drying of salami. The room should not be overloaded as there must be a uniform air flow all the time for optimal drying and to prevent growth of mold. If the fermentation or drying room is only partially filled, parameters such as temperature, RH, and air velocity have to be adjusted to avoid case hardening. If the fermentation room is only half full, case hardening occurs more easily as there is a greater volume of air available to remove moisture faster from a smaller surface area of salami. A standardized fermentation and drying program can only be followed if the room is always filled to similar levels and the product itself always has similar particles sizes and is filled into casings of similar diameters. As soon as one of these three factors changes significantly (degree of room loading, particle size of meat and fat materials, diameter of casing), the fermentation program has to be adjusted. Varying levels of fat within a product caused by a change in recipe also need to be taken into account. Higher levels of fat reduce the starting Aw of a product, while a reduction in fat increases the Aw of the product.

Fermentation starts by raising the temperature and moisture level and increasing the air flow. These three different parameters vary based on the type of salami (eg, fast-, medium-fast–, or slow-fermented salami). During the first 1–6 hours of fermentation, fast- and medium-fast–fermented salami experience an RH of around 60–70%, a temperature between 16 and 22°C, and air velocity of around 0.8 m/s. The time span of 1–6 hours is known as the "conditioning time," and its length depends largely on how full the fermentation room is and the diameter of the product. When the room is full and large-diameter casings (eg, 90 mm) are used, the conditioning period can last up to 6 hours. On the other hand, if the room is only half-full and sausages filled into a 45-mm casing are placed in the room, conditioning might last only for 1–2 hours. It is virtually impossible to give exact guidelines as different fermentation rooms work "differently" and the secret additive "experience" is very important here. More specifically, sausages are equilibrated during conditioning at a reduced RH to avoid condensation on the surface of the product.

As the cold salami (around 0°C) is placed into a significantly warmer environment, condensation water (see Section 3.9 under Chapter 3: Definitions) forms on the surface of the salami and is removed during "conditioning." Conditioning is a very cost-effective way to remove moisture from a salami. There is no need for the RH to be high during conditioning, in contrary to the other stages of fermentation, given that condensation-water produces 100% moisture on the surface of the freshly filled salami anyway. The unnecessary addition of moisture would not economically be of benefit. Myoglobin would also be washed out from the surface of the salami, contributing to a weak curing color in its outside layers.

Conditioning lasts just long enough for no more condensation water to form on the product and must also come to an end before case hardening occurs. Temperatures of only 16–22°C are used, rather than 26°C or higher, so that temperature is a hurdle against microbiological growth.

The reduction of free water also reduces enzyme activity and this can present problems, especially in small-diameter products. If small-diameter products (20–28 mm) are dried too fast during the first stage of fermentation, the Aw in the outside layers of the product is reduced to a level, which is too low for *Lactobacillus* spp. to act (Aw below 0.95) and no, or insufficient, lactic acid is formed. The pH value on those outside layers of the product remains at around 5.5 as proper acidification never takes place. Only very little curing color develops, or possibly none at all, and the low Aw on the surface finally denatures mostly metmyoglobin and the final product is of a very poor gray color.

Therefore, when producing small-diameter products, to maintain an Aw above 0.95 for a sufficient period of time for the *Lactobacillus* spp. to work and to reduce the pH value within the outer layers of the product, thus supporting the formation of curing color, there is no conditioning phase. Fermentation of small-diameter products starts at around 93–95% humidity in order not to reduce the Aw on the surface too fast.

Once conditioning of larger products is completed, the level of RH within the chamber is raised to 90–93% and the temperature is increased to around 20–28 or 35–38°C depending on the type of starter culture applied. During this period, enough moisture has to be provided to allow starter cultures to grow and produce lactic acid but the product must also dry at the same time. The air speed in the fermentation room is high (around 0.8 m/s) to remove moisture from the surface of the product. Temperatures of 24–28°C provide the perfect climate for starter cultures to grow and, in combination with high levels of moisture, production of lactic acid is ensured. The temperature-range is wide as the temperature used depends on the speed at which the pH value should decline in the product and whether fast- or medium-fast acting starter cultures are being used. From a microbiological point, temperatures above 26–28°C are not recommended. If these temperatures are used, it is slightly risky as nitrite is therefore the only hurdle in place against microbial spoilage during this stage and the use of meat and fat material displaying a low bacteria count is critical. However, temperatures of 26–30°C are commonly used in the fermentation of fast-fermented salami to achieve a drop in pH value below 5.2 within 24 hours.

The correct level of RH in the fermentation room is based on the Aw within the salami. Generally, the level of RH in the room should be 2–5% lower than the Aw within the sausage: the difference in moisture level ensures fast drying without case hardening developing. The difference in moisture levels between the fermentation room and the sausage depends greatly on the diameter of the casing used as well as the particle size of meat and fat. Moisture can be removed faster in small-diameter sausages containing coarse particles of meat and fat than in large-diameter products exhibiting small particles of meat and fat. In small-diameter sausages, the distance between the core and the surface is significantly shorter than in large-diameter products. During the production of coarse minced/mixed salami, less protein has been activated than in a finely cut salami and water is only loosely bound. In small-diameter salami made from coarse particles of meat and fat, a difference in moisture level of 4–5% can be tolerated, while in large-diameter and finely cut salami, the difference can be only 1–3%. Especially in superfine salami, cut with the help of dry ice (CO_2), the difference is commonly only 1–2% to avoid case hardening. The formula applied to calculate the correct level of RH inside the fermentation room is (Aw of sausage−[2 to 5])×100. As an example, a finely cut freshly filled salami in a large casing might have an Aw of around 0.97 at the beginning of fermentation. As a result, the RH inside the room once the period of preconditioning is over should be 94–95%, which represents a difference in moisture levels of 2–3%. When a small-diameter casing is used and the salami is made from coarse meat and fat particles, the level of moisture inside the room can be 91–92%, which is a difference of 4–5%. This difference in moisture levels should be maintained all the time for drying to take place as quickly as possible without case hardening taking place. The products are then placed in the drying room for the final drying stage.

At the beginning of fermentation, bacteria such as *Pseudomonas* spp. and Enterobacteriaceae are regularly present in the sausage.

As acidification starts to take place, the numbers of these bacteria should decrease within 1–2 days while LAB grow rapidly and come to dominate the internal microflora. During the initial stages of fermentation, the primary hurdles against microbial spoilage are the low bacteria count of the meat and fat materials used, the presence of sufficient nitrite, and elevated levels of salt. There is also less oxygen available for the bacteria inside the salami so the reduced Eh value is another hurdle against spoilage. Nitrite inhibits microbial growth more effectively at low, or reduced, Eh values, but the main advantage of reduced Eh values is that there is less oxygen available for aerobic bacteria such as Enterobacteriaceae and therefore their growth is suppressed. At the same time, the preferred lactic acid flora, predominantly made up of members of the genus *Lactobacillus*, are at an advantage as they do not require oxygen to grow. The presence of salt and NO_2 in salami during the initial stage of fermentation favors the growth of specific, facultative anaerobic, gram-positive bacteria. At the same time it inhibits the growth of fresh meat spoilage bacteria, which are primarily aerobic. Desired bacteria such as *Lactobacillus* ssp. are generally low in numbers in fresh meat and are therefore frequently added in form of starter cultures.

After around 36–48 hours of fermentation, large amounts of lactic acid are produced. The decline in pH value becomes the next dominant hurdle regarding the microbiological stability of the product once the other hurdles such as nitrite and Eh value are less effective. During the first 3–5 days of fermentation and drying, the temperature is gradually reduced to around 18–20°C and the level of moisture within the room is reduced to around 86–88%. Air velocity is also slowed down to around 0.5 m/s. Within those 3–5 days, the Aw drops below 0.95 and the growth of *Salmonella* spp. as well as other members of the family Enterobacteriacea is inhibited. The pH value drops below 5.5 in this period of time and this is another effective hurdle against growth of Enterobacteriacea. However, *S. aureus* is not yet controlled as the production of toxin by *S. aureus* is only inhibited by an Aw at or below 0.89 or a pH value below 5.2. *Salmonella* spp. are frequently a problem especially in salami made from chicken as chicken more often contains *Salmonella* spp. than other types of meat. The drop in pH value brings about a few essential changes inside the salami in regard to color, taste, aroma, sliceability, and microbiological stability.

7.3.5.1 Impact on Color

A declining pH value speeds up the development of curing color as this process occurs at a faster rate closer to a pH value of 5.2–5.3, the optimal pH range for development of curing color. More undissociated nitrous acid (HNO_2) is present at these reduced pH levels and larger amounts of NO are produced as a result (see Section 5.3 under Chapter 5: Color in Cured Meat Products and Fresh Meat). The curing color created during the decline in pH value is stabilized once the pH value is at 5.2 and below as nitrosomyoglobin is denatured by acidification as the pH value drops below 5.2.

A pH value below 5.5 greatly influences the level of activity of the enzyme nitrate reductase and no, or very little, nitrate will be reduced to nitrite below this pH value. This can present a problem when salami is produced with nitrate and needs to be acidified quickly. Low levels of nitrite are produced when a pH value of 5.5 is reached quickly simply because there is insufficient time for the reduction of nitrate to nitrite before enzyme activity from nitrate reductase comes to an end. The result is that the product will have a poor and unstable curing color. Therefore, most salami today is produced with added nitrite, or a mix of predominantly nitrite and some nitrate and in essentially all fast- and medium-fast–fermented products nitrite is the material of choice. High amounts of oxygen are normally introduced into salami, especially during cutting and mixing. If the pH drops very rapidly, especially in small-diameter products (18–26 mm), increased levels of oxymyoglobin, rather than nitroso-yoglobin, get denatured and a pinkish color will be the result.

7.3.5.2 Impact on Taste and Aroma

Acidification changes the flavor of the product and an acidic or tangy flavor is obtained. The strength of the sour taste depends on how low the pH value falls. At a pH of around 4.5–4.7 the taste is more acidic than at a pH of around 5.0. At pH values below 5.0, the conditions are favorable for heterofermentative (heterolactic) *Lactobacillus*, and they have a selective advantage over homofermentative (homolactic) *Lb*. Besides producing large quantities of lactic acid, which is desirable, heterofermentative Lb. also produce large amounts of acetic acid, CO_2 and ethanol, which are not desirable. These unwanted metabolism byproducts give the product a vinegar like taste. The formation of CO_2 also causes pores to appear, which can be seen in the finished product. In severe cases, the casing can even burst as a result of large amounts of CO_2. When salami is acidified by using either GDL or citric acid, a typical and strong acid taste is obtained, which is very different, and not desired, to the acidic taste originating from sugars being fermented into lactic acid by starter cultures, which gives a much more pleasant and typical salami taste.

The enzyme catalase is not active at pH-levels below 5.0 and H_2O_2 produced is therefore not broken down into water and oxygen. Increased levels of H_2O_2 favor development of rancidity. They also have a negative effect on curing color as H_2O_2 is a strong oxidizing agent and can destroy the globin attached to myoglobin resulting in a green-yellow color in the finished product. A very fast decline in pH from its original levels to below 5.2 also favors the growth and activity of heterofermentative *Lb*. A pH value below 5.0 brings activity of protease to a halt and so the proteases no longer produce compounds, which contribute toward the typical salami flavor. Proteolytic enzymes break down proteins into peptides and free amino acids, which contribute to the formation of the typical slightly cheesy flavor. The acid taste within salami is slightly reduced over prolonged periods of drying. Acids are oxidized over time and therefore have less on an impact on the taste on fermented salami dried for a long period of time.

7.3.5.3 Impact on Microbiological Stability

LAB, which are mainly responsible for the formation of lactic acid and therefore for the decline in pH value, act in the water phase of the formulation. Any factor reducing the Aw below 0.95 at the beginning of fermentation, therefore, results in the formation of very little lactic acid or none at all. The curing color will then be poor and, most importantly, the pH will never be a hurdle against microbial spoilage. It can be problematic to use freeze-dried meat in the recipe as this could lower the initial Aw quickly below 0.95, thus deactivating LAB.

Ongoing acidification leads to a more microbiologically stable product as most bacteria react very sensitive to increased levels of acidity in their environment. Lactic acid obtained from fermented sugars as well as gluconic acid from GDL (as well as other acids produced in tiny amounts) are responsible for the decline in pH value. Enterobacteriacea such as *Salmonella* spp. are inhibited at/below a pH value of 5.5 and *S. aureus* does not produce toxin below a pH value of 5.2. By reaching a pH value of 5.2 and below, the product is microbiologically stable due to acidification. At reduced pH levels, nitrite also becomes a more effective hurdle against bacterial growth. The acid produced when starter cultures ferment sugar is more effective in inhibiting the growth of *Salmonella* spp. than the gluconic acid produced from GDL even when the final pH in the salami is the same. In small-diameter products, when the Aw in the outer layer of the sausage is reduced too fast to below 0.95 and when the change in pH is due to starter cultures fermenting sugars, *Lactobacillus* spp. is inhibited at the beginning of fermentation. Therefore, acidification never takes place and the product is at great risk of microbiological spoilage given that the outer layers will have a high Aw for a long period of time. *S. aureus* can grow in those nonacidified areas of the product: it is only inhibited by an Aw at or below 0.89, and it takes time to reduce the Aw in the nonacidified areas below that level.

7.3.5.4 Impact on Sliceability

Another important aspect of a declining pH value is that a move toward the IEP increases the possibility of removing moisture from the product as WHC is reduced at lower pH levels close to the IEP. As a result, high amounts of moisture can be removed from the salami during the decline in pH value from around 5.7 to 5.2. The viscous sol, obtained during cutting in the bowl-cutter or during mixing, is transformed into a solid gel at a pH of 5.2. At a pH value of 5.2, activated protein within the sol coagulates due to acidification and the salami becomes sliceable. Subsequent drying increasing sliceability even further. Crumbly products with poor slice coherency are commonly due to insufficient levels of sol forming during cutting, or the temperature during mixing of the sausage mass is too cold with water being present as ice. The latter results in little sol transforming into a gel via acidification.

The time taken for the pH value drop below 5.2, or below 5.0, depends on the type of sugars added, the speed at which the starter cultures act, and the temperature during the first 48 hours of fermentation. A combination of fast-acting starter cultures, glucose (which is directly fermented into lactic acid), and elevated temperatures (around 26–28°C) cause the pH value to drop quickly and pH values of or below 5.2 can be obtained within 24–36 hours. The same takes place when GDL is added as elevated temperatures speed up the formation of gluconic acid. However, a rapid decline in pH value (below 4.8 within 24 hours) when GDL is broken down to gluconic acid quickly causes the formation of large amounts of CO_2, which can lead to gassing in the packed salami. The final pH of the product depends on the amount of sugar or GDL added as greater amounts lead to higher levels of lactic or gluconic acid. However, the acidification that takes place owing to the action of starter cultures can be stopped once the desired pH value is reached by reducing the temperature within the fermentation room to around 10–12°C as those temperatures are too low for LAB to ferment remaining sugar in lactic acid. The residual sugar is then used for color and flavor development rather than acidification. Acidification induced by GDL or citric acid cannot be stopped even at reduced temperatures as it is a chemical process and as long as free water and GDL are available, GDL turns into gluconic acid regardless of the temperature at which the reaction is taking place.

After fermentation of around 36–76 hours the salami has a pH of 5.2 or below. The pH value is the dominant hurdle against microbiological spoilage and the product is microbiologically stable. Generally, the pH drops further to levels between 4.6 and 4.9 to have some "safety buffer" given that a pH value of 5.2 is the maximum. A pH value of 4.6 is sour, which is usually not pleasing to the consumer. However, from a microbiological point these low pH values are very safe. In countries such as the United States, a summer sausage is produced, which has a pH in the range of 4.6–4.8 or even lower, and this product is accepted by consumers. Around 4–5% in weight can be lost during each day of fermentation without obtaining case hardening when modern fermentation and drying technology is applied properly.

7.3.5.5 Smoking During Fermentation

It is very common to smoke raw fermented sausage to create the typical smoked color and flavor. Smoking takes place at 20–25°C. The application of smoke helps to prevent growth of molds and a slight antioxidative effect is even seen on the surface as phenols, which are present in smoke, act as an antioxidant by deactivating free fatty acid radicals. Products produced worldwide have different colors from very light to dark black because of the different degrees of smoking. Most commonly, a slight golden brown is desired. Salami is smoked for the very first time during fermentation once the red curing color is fully developed and stabilized. Salami is not smoked within the first 36–48 hours given that components within smoke, such as phenols and other acids, have a negative impact on the development of curing color, especially on the surface of the product. In fast- and medium-fast–fermented salami, smoke is generally applied after around 36–48 hours of fermentation. At this stage, the pH value has dropped to 5.2 or below and a stable curing color has developed (at a pH value of 5.2, and below, nitrosomyoglobin has been denatured and therefore stabilized via the impact of acidification). The reduced level of humidity means that the casing is dry and therefore ready to be smoked. Smoke applied too early during fermentation results in a brown-colored product. No mold is seen on the product after 36–48 hours if the climate within the fermentation room is controlled properly. If the casing is too wet during the initial stages of fermentation and smoke is applied at this stage, a spotty, dark, and uneven color will be seen on the product.

Cold smoke is not carried out as one long continuous process and is applied in intervals lasting 1–3 hours several times per day for 2–3 days or as often as necessary until the desired smoke color is reached. Control of RH and air circulation is critical so that the casing is uniformly dried before smoking. During fermentation, as explained earlier, the RH in the chamber is slightly lower as the Aw within the salami. As a result, the casing generally has just the right level of moisture after 36–48 hours of fermentation to take up smoke.

The small bundle of casing on the sausage after the clip can be used to check the RH in the room. This bundle should feel moist but not soaking-wet or paper-dry. If this bundle feels soaking-wet, the RH is normally too high, whereas the RH is generally too low if the bundle feels paper-dry and almost brittle.

In large-scale production, all production steps such as fermentation, smoking, and drying takes place in one and the same room. A batch of salami remains in the same room until the desired loss in weight is obtained.

Based on the microbiological stability of the product at a pH of 5.2 and below, a fast-fermented salami is usually ready to be dried after around 36–48 hours. A pH of 5.2 and below is reached in a medium-fast–fermented product after around 76–96 hours.

7.3.5.6 Flavor in Salami

The flavor within salami depends to a large extent on the time given to the product to develop the typical salami flavor as well as the type of spices added to the product in first place. To obtain the typical mellow and tangy salami flavor, the enzymes such as proteases and lipases need time to act. During flavor development, proteases such as calpain and cathepsins break down proteins into peptides as well as free amino acids. Bacterial proteolysis, though, has little impact on the typical salami flavor. Free amino acids such as valine, leucine, taurine, and glutamine are produced through proteolysis and they contribute to flavor. Lipases break down fat into free fatty acids, which also contribute strongly to flavor.

These enzyme-based reactions are the major source of flavor in slow-fermented and long-dried salami. Over 250 different flavors compounds are now known to be present in slow-fermented salami but no single determining flavor has yet been found. Even rancidity is desired in slow-fermented salami to a small degree as it contributes to flavor. Compounds such as aldehydes and ketones contribute to the "desired" rancid flavor.

The factor "time" does not play a role in flavor development in fast fermented salami as those products are commonly sold within 5–21 days of production, depending on the diameter of the casing used. The flavor of products dried for a short amount of time is predominantly determined by their internal acidity as well as the addition of spices. Their acidity either comes from lactic acid or GDL, or a combination of both, and it gives the product a sour and tangy flavor. In the production of fast-fermented salami, there is no time for any significant enzyme activity such as proteolysis or lipolysis to contribute to flavor development. In the manufacture of medium-fast–fermented salami, there is a small amount of time available for flavor development given that these products are sold around 3–5 weeks after production. The final flavor in these products is determined by a combination of internal acidification, the spices introduced and the product proteolysis and lipolysis to some degree.

Smoking (see Section 4.15 under Chapter 4: Additives) inhibits the growth of mold as phenols and carbonyls are present, but it also has a major impact on the flavor of the product. The presence of selected molds on the surface of the product also contributes to flavor as they break down protein and fat and penetrate into the sausage. Some lactic acid is metabolized by molds, which causes a raise in pH during extended drying periods.

Rancidity is generally an unwanted flavor in salami. Some degree of rancidity, though, is desired in products dried for a long time. These products are primarily appreciated by salami-lovers and commonly not by the everyday consumer. To delay rancidity, fat with a low level of unsaturated fatty acids should be processed as a raw material and this fat should not be rancid at all at the point of processing. Filling a product under vacuum also reduces the level of oxygen inside the sausage mass, thus slowing down the development of rancidity. In addition, nitrite added to the product slows down the development of rancidity as NO binds the iron present in the mass (unbound heavy metal ions, such as iron, and quicker oxidation, causing rancidity). The presence of catalase, predominantly produced by members of the family Micrococcaceae, breaks down

H_2O_2 into water and oxygen, delaying rancidity. The application of smoke, as well as antioxidants, can also slow down rancidity by deactivating free radicals.

7.3.5.7 Salami Manufacture Without a Fermentation Room

There are some special methods for a fast reduction in Aw in salami to be used when proper fermentation rooms are not available. These techniques are occasionally practiced in a small-scale operation and products filled into casings up to around 60-mm diameter are treated this way.

7.3.5.7.1 Alternating Program

After filling, the salami is kept in condition of low RH (60–70%) and fast air speed for around 10–16 hours at 22–26°C. It is kept to a point shortly before case hardening would be occur. At this point, RH is increased to around 92–94% for several hours. Afterward, RH is lowered again and high air speed is applied again for around 8 hours. This treatment reduces the Aw quickly below 0.95 and more gentle and gradual drying can follow. This method of reducing the Aw quickly is occasionally used in the manufacture of slow-fermented salami where the pH value never acts as a hurdle. Such a quick reduction in Aw during the initial stage of fermentation requires sound knowledge of the process and experience as once case hardening occurs, drying afterward would be greatly reduced overall. This treatment therefore has to be fine-tuned and parameters such as different diameter casings and the size of meat and fat particles vary the conditions required.

7.3.5.7.2 Salt Method

Filled salami, containing around 22 g of salt/kg, is placed in tubs or containers and covered in pure salt for around 3 days under chilled conditions. The salt, which is highly hydroscopic, is changed every day. Water penetrates from the sausage into the salt as the concentration of salt is much higher within the surrounding salt itself than inside the sausage. This results in a fast reduction in Aw within the sausage. This method is sometimes used in the manufacture of slow-fermented salami. No, or very little, acidification takes place given that LAB, such as *Lactobacillus* spp., do not have sufficient moisture at an Aw below 0.95 to ferment the sugars present in meat into lactic acid. The maximum diameter of products treated this way is around 60 mm as the reduction in Aw in larger-diameter products would require a significantly longer time.

7.3.5.7.3 Brine Method

Filled salami is placed in brine containing 10–14% salt and the high concentration of salt within the brine lowers the Aw of the sausage fairly quickly. This method has the disadvantage that some myoglobin is washed out during soaking and less curing color is seen in the final product.

7.3.5.7.4 Vacuum Drying

The filled salami is placed under vacuum at around 60% vacuum. A full vacuum cannot be applied as the clip on the casing would move or the casing burst. After around 2–5 hours the vacuum is released and another 3–6 hours later, a slight vacuum is applied again. The vacuum applied opens up capillaries and moisture can be removed during the nonvacuum periods efficiently before the product is placed under a vacuum again.

All these alternative methods can only be used in small factories and do not suit large operations, their advantage being that most of those methods can be used without having to have a proper fermentation room as the Aw is quickly reduced below 0.95, inhibiting the growth of *Salmonella* spp. Further drying to a lower Aw also eventually stabilizes the product against *S. aureus*. Safe, secure, and efficient fermentation as well as drying of salami without proper fermentation and drying room is hard to achieve. Not being able to control climatic conditions such as RH, temperature, and air velocity either leads to the formation of case hardening if large amounts of moisture are removed from the surface of the product too fast or to noneffective drying and the growth of unwanted mold if the RH is too high. Therefore, all medium- and large-scale operations have proper fermentation and drying rooms in place.

Salami, produced with GDL, can be made on a small scale without a proper fermentation and drying room. The filled product is placed in an area with a temperature of between 22 and 28°C for around 36–48 hours. The sausage is showered every 3–4 hours for several minutes only to maintain a moist surface to avoid case hardening, and the salami should be exposed to moderate air speed. After 36–48 hours, the pH value has dropped below 5.2, making the product microbiological stable. The salami is cold smoked several times and then placed in an area with slow air speed (around 0.1–0.3 m/s) for further drying. The utilization of GDL is recommended within this method given that the conversion of GDL into gluconic acid is a chemical process, as opposed to a microbiological process such as applying starter cultures and sugars, and is a safe way of acidification. However, starter cultures can be used as well but a chemical conversion of a material into acid is a more secure way compared with a biochemical method where living organisms have to ferment a substance to obtain the desired acidification. When working with GDL, at least 8 g should be applied per 1 kg of sausage mass so that the pH value drops to a safe level of around 4.8. In extreme cases, the freshly filled product containing GDL or starter cultures and sugars is placed for 24 hours in 24–28°C warm water, meaning that the product is exposed to an RH of 100%. The water is at the temperature required for starter cultures to ferment sugars into lactic acid as well as for GDL to hydrolyze into gluconic acid. A small degree of myoglobin will be washed out during soaking, but this is not of concern as the salami made in this way is not intended to be a "top-quality product." The intention is rather to make salami in a simple way. The product is removed from the water and hung at 24–28°C and under conditions of high humidity for 1 more day. If the level of humidity cannot be controlled, the product is showered every 2–3 hours for a few minutes to maintain a high level of moisture on the surface of the product. The drop in pH value below 5.2 will be achieved after 36–48 hours and the product is stable as a result. Smoking and some degree of drying complete the process.

7.3.5.8 Fast-Fermented Salami

A large amount of salami today is produced quickly and used as pizza topping or for other purposes. Fast-fermented salamis are most often stabilized by pH value only and water activity never comes into play as a hurdle against microbiological spoilage. Generally, these products are sold within 5–10 days of manufacture and small- to medium-sized–diameter casing are used with 50–55 mm being the largest diameter used. For the purposes of acidification, fast-acting starter cultures in combination with glucose and/or GDL or citric acid are the materials of choice. Some products are made with GDL only. The pH value of fast fermented salami is around 4.6–4.8 in the finished product, which makes it microbiologically stable. Commonly, the addition of around 10 g of glucose, or a combination of 5 g glucose (and starter cultures) and 5 g of GDL per 1 kg of salami reduces the pH value by 1 full pH unit (from 5.7 to 4.7, as an example). Alternatively around 10–12 g of GDL is added per 1 kg of sausage mass, which also reduces the pH value by around 1 pH unit. Nitrite, and no nitrate, is added as the time required for nitrate to be reduced to nitrite is not available. Nitrite also has to act as a hurdle against microbiological spoilage during the initial stage of fermentation. Starter cultures commonly used in the manufacture of fast-fermented salami contain predominantly *Lactobacillus* spp. and *Pediococcus* spp. but occasionally also small amounts of *Micrococcus* spp. and *Staphylococcus* spp. for their contribution to color and flavor within the product. Flavor in fast fermented salami is greatly determined by the presence of lactic acid and the low pH value give a sour and tangy flavor. Very little of the typical "buttery" salami flavor is present given that there is insufficient time for enzymes to break down proteins and fat. As a result, the spices added to the product in combination with its acidic/tangy note determine the flavor. Fermentation usually takes place at temperatures between 26 and 30°C as this enables fermentation to get going quickly. The final Aw of the product is around 0.92–0.94, which represents a loss in weight between 10% and 20%. Large volumes of fast-fermented salami are produced in a 5–6 day cycle, and product produced on Monday is sold (or packed) on Friday or Saturday of the same week. Some countries' food standards insist that the pH value within fast-fermented raw fermented salami must drop below 5.2 within 48 hours (Figs. 7.2–7.4, Table 7.1).

7.3.5.9 Medium-Fast–Fermented Salami

Medium-fast–fermented salami is usually acidified by a mixture of glucose and GDL, or by the addition of sugars together only with starter cultures. If sugars

Fermented Salami: Non–Heat Treated **Chapter | 7** **155**

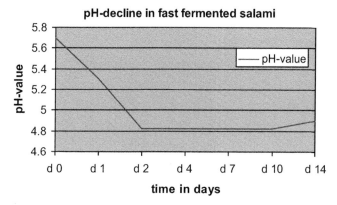

FIGURE 7.2 The decline in pH value of fast-fermented salami.

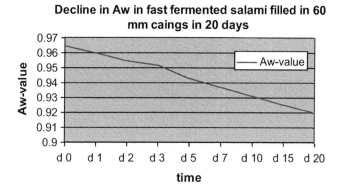

FIGURE 7.3 The decline in Aw in fast-fermented salami.

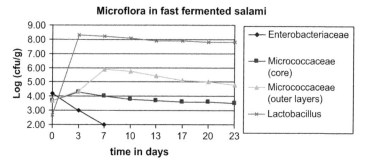

FIGURE 7.4 The microflora in fast-fermented salami.

TABLE 7.1 Typical Fermentation Program for Fast-Fermented Salami Filled in a 45-mm Fibrous Casing

Time	Temperature (°C)	Humidity RH	Smoke
1–6 hours	24–26	50–60%	No
Day 1	24–26	90–92%	No
Day 2	20	90–92%	No
Day 3	20	88–90%	Yes
Day 4	18	85–87%	Yes
Day 5	18	83–85%	No
Day 6	16	80–82%	Yes
Day 7 onwards	14	72–75%	No

are used, the majority of the sugar added is glucose as this causes the pH value to decline fast below 5.2 but other sugars such as lactose are added as well. Combinations of glucose, lactose, maltose, and some GDL can also be added. The starter cultures chosen do not work as fast as the starter cultures applied to fast-fermented salami. Temperatures of 22–24°C are normally applied during the initial stage of fermentation. The addition of 4 g of glucose and around 6 g of lactose normally lowers the pH to around 5.1. Medium-fast–fermented salami, depending on the diameter of casing used and desired weight loss, are generally sold after 14–28 days after production. Nitrite, and no nitrate, is added and nitrite becomes the first hurdle against microbiological spoilage. A declining pH value is the next hurdle and levels of 4.8–5.0 are commonly aimed for. When the pH value remains below 5.2, Aw is established as the next hurdle. Some medium-fast–fermented salami sold are stabilized against microbiological growth only by their pH value below 5.2, some are stabilized by an Aw below 0.89, and others are sold with a pH value of around 5.0–5.1 and an Aw of around 0.93. At these pH levels, growth of *S. aureus* and *Salmonella* spp. is inhibited. An Aw of 0.93 is an additional hurdle against *Salmonella* spp. given that an Aw above 0.95 is required for *Salmonella* spp. to grow. Care has to be taken in medium-fast–fermented salami that the pH value does not rise above 5.2 as long as the Aw is not at or below 0.89 as no hurdle would be in place against *St. aureus* and this would be a serious health risk. An Aw of 0.89 or below is required for *St. aureus* not to produce its toxin. The flavor of medium-fast–fermented salami is due to acidification, addition of spices, and some degree of proteolysis and lipolysis, if the product is dried for around 4–5 weeks. Generally, medium-fast–fermented salami have a stronger salami flavor compared with fast-fermented salami (Fig. 7.5).

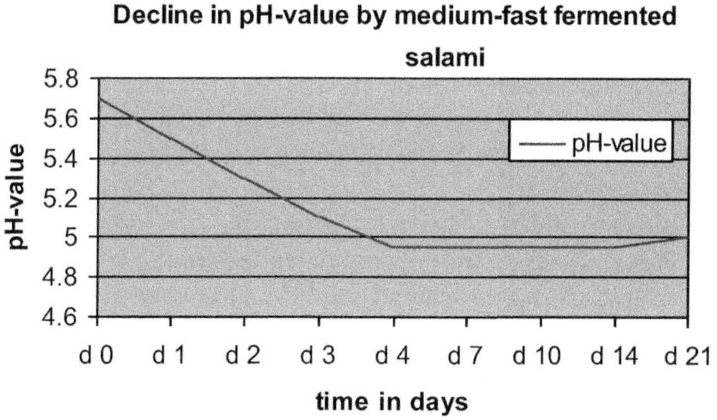

Decline in pH-value by medium-fast fermented salami

[Chart showing pH-value declining from ~5.7 at d0 to ~4.9 by d7, remaining stable through d21, with x-axis: d0, d1, d2, d3, d4, d7, d10, d14, d21 and y-axis: pH-value from 4.6 to 5.8]

FIGURE 7.5 Decline in pH value in medium-fast–fermented salami.

7.3.5.10 Slow-Fermented Salami

Slow-fermented salami is the "classic" type of salami but it cannot be produced in many countries as their food standards call for a drop in pH value in a salami within the first 48 hours of fermentation to 5.2 or below. "Classic" salami never experiences this drop in pH value and Aw is instead the main hurdle in this product against microbial growth: the pH value never becomes one of the main hurdles. To produce a safe slow-fermented salami, as with all other types of fermented salami, the meat and fat raw materials must have a low bacteria count.

Small amounts of sugar are sometimes added to support the natural acidification achieved by the *Lactobacillus* spp. present in meat. Starter cultures, or more specifically protective (or competitive) cultures, are also frequently added to inhibit growth of bacteria that are naturally present but undesirable (such as *Salmonella* spp.) until an Aw of 0.95 is reached. Protective cultures, though, are not added for the purpose of acidification as such. Protective cultures frequently contain *Micrococcus* spp. and *Staphylococcus* spp. These bacteria are generally added for their contribution to development of a strong curing color and excellent flavor, rather than for the purpose of acidification. Naturally present *Lactobacillus* spp. ferment added sugars, generally between 2 and 4 g/kg of sausage mass, and in case the pH value tends to drop to levels below 5.3–5.2, the temperature inside the fermentation room is reduced to 10–12°C as the lactic acid bacteria stop fermenting sugars into lactic at such low temperatures. The remaining sugar is then used for the development of color and flavor.

As a result of natural acidification, the pH value obtained is by around 5.2–5.3, which is another hurdle against bacteria such as *Salmonella* spp. and *S. aureus* is also partly inhibited. It is not uncommon, though, for the pH value within the product to increase during the first 1–2 days of fermentation by around 0.1–0.2 pH unit due to enzyme activity until natural acidification starts to take place. Fermentation in naturally fermented salami is caused by the lactic acid bacteria naturally present in meat. Numbers of these bacteria can increase to 10^5-10^6/g within 3–5 days of fermentation while other bacteria such as *Pseudomonas* spp. decrease sharply in numbers. Experience shows that the natural acidification is initiated first by some members of the family Enterobacteriaceae, followed by *Enterococcus* spp., *Pediococcus* spp., *Streptococcus* spp. And, last, *Lactobacillus* spp. *Streptococcus* spp. ferment glucose into lactic acid at temperatures above 14°C.

The pH value remains at around 5.2–5.3 until the Aw is at a suitable level to be a hurdle. The pH value should not drop quickly below 5.5 as nitrate-reducing bacteria, such as *Micrococcus* spp., do not reduce nitrate to nitrite at pH-levels below 5.5, which is a problem when nitrate is added in significant levels instead of nitrite. A fair amount of nitrite is always oxidized to nitrate anyway, and this has to be reduced to nitrite again. If the pH value drops quickly, and nitrate is present, poor curing color would is the result as little nitrite would be present.

During initial drying, the Aw commonly drops below 0.95 after around 4–6 days, which provides sufficient time for natural acidification due to the action of *Lactobacillus* spp. and curing color can develop as well. Natural acidification supports the formation of nitrous acid (HNO_2) and elevated levels of undissociated HNO_2 result in higher levels of NO, which forms the stable curing color (see Section 5.3 under Chapter 5: Color in Cured Meat Products and Fresh Meat). Drying of slow fermented salami must not happen too quickly within the initial stage of fermentation as a reduction in Aw below 0.95 in the outer layers of the product would not provide sufficient water for *Lactobacillus* spp. to act. Poor curing color would be seen in the finished product as there would be little undissociated HNO_2.

During this initial stage of fermentation, the low bacteria count of the meat and fat used, the addition of salt (at least 26 g/kg), nitrite, and often some nitrate, and fermenting temperatures of only around 12–16°C are the main hurdles against microbiological spoilage until an Aw below 0.95 is achieved. At an Aw below 0.95 growth of Enterobacteriaceae such as *Salmonella* spp. is inhibited. Growth of Enterobacteriaceae is inhibited once the loss in weight is around 15% resulting in an Aw of or below 0.95 as well as obtaining increased levels of salt at the same time due to the loss in weight.

If the product is smoked, smoking takes place after around 48 hours to prevent growth of unwanted mold. A large percentage of slow-fermented salami have noble mold on their surface. If this is the case, the salami is normally not smoked at all or is smoked slightly before the noble mold is applied. Application of mold happens via spraying after around 2–3 days. In the manufacture of nonsmoked slow fermented salami, mold is also frequently applied by dipping the freshly filled salami into a suitable solution.

Slow-fermented salami is sold after 6 weeks to 5 months and the time of sale depends greatly on the diameter of the casing used. Some large-diameter products are even dried for 8 months to a year. The sol obtained during mixing of the salami mass transforms into a gel in slow fermented salami due to the presence of a low Aw in conjunction with high levels of salt. This gel is never fully acidified. This is in contrast to fast- or medium-fast–fermented salami, in, which the transformation of sol into gel is achieved via the impact of acidification (pH value below 5.2). The final Aw in the product is normally between 0.82 and 0.88 and pH values of 6.0–6.4 are frequently seen, which do not present problem in terms of microbial growth given that an Aw below 0.89 stabilizes the product. Commonly, a mix of nitrite and nitrate is applied for the development of the curing color and the application of nitrate ensures a stable color even after months of drying. To produce a salami by this method safely requires a lot of experience and knowledge of the materials and processes used.

Given that this product is never fully acidified, the flavor of the nonacidified product is very distinctive, it is the slightly sweet and tangy-buttery flavor typically associated with salami. The typical salami flavor is predominantly due to proteolysis (free amino acids) as well as lipolysis. Enzymatic alkaline metabolic byproducts lead to an increase in pH value to levels above 6.0 in the finished product and therefore a slight degree of rancidity is frequently part of the flavor as well. The lipolytic activity of lipase from lactic acid bacteria is significantly less than that of *Micrococcus* spp. and lipase from molds are also of importance in slow fermented salami. After lipolysis, fewer triglycerides are present and a higher degree of diglycerides and free fatty acids are obtained. The flavor in salami dried for a long period of time can due to up to 300 different compounds. The flavor is typically a result of slight acidity, proteolysis and lipolysis and the presence of aldehydes and ketones and many other compounds.

Caseinate is regularly added to slow fermented salami at 5–10 g/kg. Caseinate binds water to some degree and drying is slightly slowed down as a result of its presence. A prolonged period of drying gives enzymes the opportunity to create higher levels of free amino acids and free fatty acids, which contribute to flavor. The flavor from caseinate itself naturally matches with the flavor of salami dried for a long period of time. A degree of bitterness can be seen in nonacidified salami dried for a long period of time as a result of the death of proteolytic enzymes.

Case hardening has to be avoided during the early stage of fermentation and drying within all different methods of fermentation because the Aw cannot be reduced to a level below 0.95 and a high water activity supports accelerated activity of protease. Large amounts of highly alkaline byproducts are formed by protease causing a rapid raise in pH value. This results in uncontrolled fermentation as the natural lactic-acid flora never have a chance to acidify the product to levels around 5.3. If the salami has a high pH values for a prolonged period of time, there is no hurdle in place against *Salmonella* spp. and *S. aureus*. Both can therefore grow uncontrolled and the product is unsafe (Fig. 7.6).

Even though the final pH values reached are similar, the difference is that in a controlled fermentation a low Aw stabilizes the product while obtaining a high pH value as a result of prolonged drying. In an uncontrolled fermentation, the pH value keeps rising right from the beginning without ever declining leading to immediate spoilage.

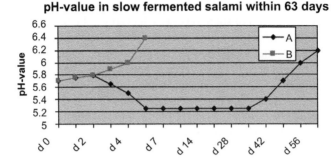

FIGURE 7.6 pH value of controlled (A) and uncontrolled (B) fermentation in slow fermented salami.

If case hardening occurs in the late (or later) stages of drying when the product is microbiologically safe and at an Aw below 0.89, this is generally not a problem. The only difficulty is that case hardening is not appealing to the customer. There is greater risk of case hardening in large-diameter slow-fermented salami compared with in small-diameter products just as in the manufacture of fast- and medium-fast–fermented salami. Less water can be removed from the surface of the product within a certain period of time given that water has to travel a longer distance from core to the surface. The extended distance slows down the speed at which water travels. This explains why drying has to take place at a slower rate in large-diameter products compared with small-diameter products (Figs. 7.7 and 7.8).

As in the manufacture of other types of salami, smearing of fat must be avoided at any stage of the process. Fat smearing would cause capillaries to be closed and therefore it would not be possible to use Aw as a hurdle since drying, and therefore the removal of moisture, would be badly affected. Given that the pH value never becomes an effective hurdle against microbial growth in slow-fermented salami, it is detrimental if case hardening occurs as a result of smearing. This could give unwanted bacteria an advantage over the desired microflora and microbiological spoilage could be the result.

7.3.5.11 Salami With Noble Mold

In places such as France and Italy, a large percentage of all salami produced has mold on its surface. The presence of this mold is intentional and it is applied to the surface of the product either in a dipping or spraying process. Genera of *Penicllium* are generally chosen. Few salamis have yeasts such as *Debaryomyces hansenii* or *Candida famata* on the surface. Yeasts are sometimes introduced directly into the sausage mass for flavoring purposes and *Debaryomyces hansenii* at levels of 10^6–10^7/g of sausage is generally chosen. Mold is applied either to medium- or slow-fermented products: these products are generally fermented and dried for at least for 4–6 weeks. The color of mold present on the surface

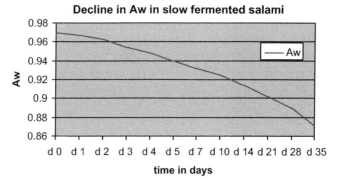

FIGURE 7.7 Decline in Aw in slow-fermented salami.

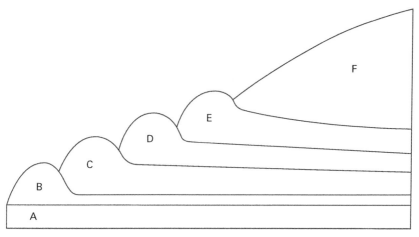

FIGURE 7.8 Sequence of hurdles in slow-fermented salami where finally an Aw below 0.89 stabilizes the product. One hurdle is constantly in place right from the beginning of the process. A, meat and fat material showing a low bacteria count; B, the presence of nitrite; C, reduction in Eh value; D, competitive flora (lactic acid bacteria); E, decline in pH value; F, reduction in Aw as a result of drying therefore strengthening (increasing) the Aw as a hurdle.

of the product should be white or off-white, rather than green, yellow, blue, or black. All unwanted molds are usually of one or more of these nonwhite colors.

The desired mold should cover the product surface evenly and fully. If growth of mold is not taking place at the desired speed, a slight rise in temperature as well as RH often provides the small change in climate required for optimal growth. A mold-dipping solution contains around 10^6–10^7 cells/mL of water. When the solution is applied by spraying it onto to the surface the product, rather than by dipping the filled product into the solution, it should not be too cold. The surface of the casing should also not be too moist (wet) as the mold spores cannot grip onto wet surfaces. Generally, mold solution is sprayed on after around 2 days of fermentation once all condensation water is removed and drying has already taken place to some degree. Slightly smoked sausages can only be inoculated, typically via spraying, once volatile smoke components such as phenols and formaldehyde, are no longer present on the surface of the product, as they would interfere badly with the growth of mold. More specifically, around 24–48 hours should pass between a light application of smoke and the subsequent spraying of the product. The majority of mold-covered salamis are never smoked, but a touch of cold smoke at around 20–25°C after around 2 days of fermentation helps to suppress any unwanted mold.

The presence of mold on the surface of the product stops, or delays, case hardening and contributes positively to the flavor of the product as mold generally contains proteases and lipases. Mold also protects the product, or the surface of the product, from the impact of light and oxygen and therefore stabilizes the color and delays rancidity. Before the product is packed and sold, most of the mold present on the surface is brushed off as some mold can grow to a substantial length. If an uneven layer of mold is obtained during drying, the brushed salami can be rolled in talcum or flour so that it looks like the mold covers the surface of the product more evenly.

7.3.6 Drying of Salami

In drying the product, the aim is to lose the desired amount of moisture in the shortest possible time without case hardening occurring making the product ready for sale. As during fermentation, the RH in the drying room should be 2–5% lower as the Aw inside the salami to ensure efficient drying without obtaining care hardening. The process of drying, and therefore the reduction in Aw, has a significant impact on taste, flavor, texture, and color of the product. It is hard to define the point at, which fermentation comes to an end and the process of drying starts because a loss in weight of the product occurs right from the beginning of the fermentation process. Generally, the drying phase starts once microbiological stability is obtained (when the salami's pH value is stabilized), which is the case once the pH has reached 5.2. In a nonacidified salami, which is only slow fermented and dried, drying basically starts right from the beginning of the fermentation process even though the time in, which the pH declines from around 5.7–5.3 could be seen as the fermentation phase.

It would not be possible to dry salami if salt were not added to the product. Given that the RH in the drying room is constantly kept lower than the Aw of the salami, there is a difference in vapor pressure causing the removal of moisture from the outside layers of the salami. As a result, the concentration of salt is enhanced in the outside layers of the product due to a decreased Aw. The difference in moisture level between the core and the surface of the product must be balanced out and therefore water diffuses from the core of the product toward the surface to equalize the levels of salt and Aw (osmosis). The outer layers of the salami always have a lower level of moisture compared the core of the salami due to the constant reduction in RH in the drying room and water continuously diffuses from the core toward the surface. Consequently, the core of the salami dries before the outer layers. The speed of evaporation of water from the surface of the product must be suitable for the speed of water penetrating from the core toward the surface. When moisture is removed faster from the surface of the product at a faster rate than that at, which it diffuses from the core toward the surface, case hardening will be the result.

The size of the meat and fat particles present within the product, diameter of casing used, fat content of the product, and air speed (air velocity) determine the maximum difference in vapor pressure between the sausage (Aw) and the atmosphere in the drying room (RH) and therefore the amount of moisture removed from the surface of the product within a certain period of time.

Generally, parameters such as elevated temperatures, high air velocity, and low RH speed up the removal of moisture from the surface of the product. Reduced temperatures, reduced air velocity and elevated levels of RH slow down removal of moisture. These parameters have to be adjusted to dry the product as fast as possible without case hardening occurring. The amount of moisture to be removed from the surface of the product filled into a large-diameter casing is less than that to be removed from a small-diameter product as the distance from the core of the product to the surface is longer in a large-diameter product. Finely cut products with a particle size of 0.8–3.0 mm must be dried more slowly than products made from coarse meat and fat particles even when filled in the same-diameter casing. This is because the migrating moisture encounters meat and fat particles with a significantly larger surface area on its way from the core to the surface. Due to the presence of many small particles in the product, the stream of migrating moisture is redirected much more often than is the case in a coarse product thus increasing the distance moisture has to travel until it finally reaches the surface of the product (Fig. 7.9).

During drying, the temperature within the drying room is gradually reduced to 12–14°C, the RH is gradually lowered to around 72–75% and air velocity is ultimately reduced to 0.1 m/s. Air flow must never come to a complete standstill as mold could develop quickly. Temperatures of around 12–14°C are also used as

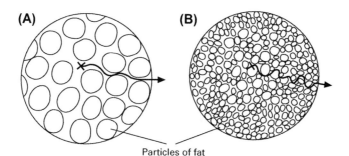

Particles of fat

FIGURE 7.9 Distance and redirection water faces on its way from the core to the surface in coarse (A) or finely cut (B) products.

they do not support the growth of mold. Through drying and the subsequent lack of free moisture, microbial growth as well as enzyme activity is largely inhibited. However, enzymes such as protease and lipase are still active during the drying process, which is desired for flavor development inside the product. The speed of drying has to be adjusted to the diameter of the casing as well as the particle size of meat and fat shown inside the product. As during fermentation, moisture can be removed in the later stage of drying at a faster rate from small-diameter products with coarse-sized meat and fat particles than from large-diameter products made from small-sized meat and fat particles. As a result, the fermentation and drying program has to be fine-tuned according to these parameters. If a small-diameter product is fermented and dried using a program designed for large-diameter products, slow and uneconomical drying will be the result and the Aw will remain too high for an extended period of time during fermentation. A higher degree of acidification will take place as *Lactobacillus* spp. can ferment sugars for a longer period of time into lactic acid and mold will have a chance to grow as well. A similar principle applies during drying. If a product is not dried at the required speed, the desired firmness (or loss in weight) will not be obtained so soon and every day of extended drying due to insufficient removal of moisture is very costly. The risk of mold growing due to ineffective drying is greatly enhanced as well. To optimize drying in such a scenario, RH can be reduced and air velocity can be increased, but care must be taken that these adjustments do not cause case hardening.

On the other hand, if a fermentation and drying program based on a small-diameter product is applied to a large-diameter product, case hardening will occur and microbiologically unstable as well as visually unattractive products are the result. More moisture is removed from the surface of the product than is the amount penetrating from the core of the product toward the surface. Therefore, the stream of moisture is interrupted and case hardening is the result. To avoid case hardening in such a scenario and to remove less moisture from the surface of the product within a given period of time, air velocity can be reduced, the level of RH can be increased, or a combination of both.

Drying rooms require uniform temperature, humidity and air velocity. Generally, obtaining an even climate within the room and making sure that the RH and air velocity are the same in all its corners is more difficult in larger-sized drying rooms. The longer the distance the air has to travel the more difficult is becomes to maintain a uniform air flow and level of RH. A uniform climate is essential to achieve the fastest possible drying time without case hardening and mold-growth occurring. Commonly, some of the salami placed in a room exhibits case hardening while others products display growth of mold. These are signs of an uneven airflow and an uneven climate overall. During production on a small scale it is possible to counteract those problems by shifting trollies around the room to overcome those shortfalls. On a larger scale, this option of solving the problem of an uneven climate is not possible. When there is uneven air flow in the drying-room (which is a common problem especially in large-sized rooms), or products of different diameters and/or particle sizes are in the same drying-room at the same time, the speed of drying is adjusted to the product most liable to case hardening (that from, which the smallest amount of moisture can be removed within a given period of time before case hardening occurs; usually large-diameter products with small sized meat and fat particles). By doing so, the cases of all the other products, which could be dried faster, do not harden either but the drying of products in the room from, which removal of moisture could take place at a faster rate (small-diameter and/or coarse meat and fat particles) is less economical. Inefficient removal of moisture favors the growth of unwanted mold and therefore products of the same, or similar, diameter should be dried whenever possible at the same time in the same room.

During drying, if growth of mold is seen, it is generally more effective to reduce the RH in the room slightly and slow down the air velocity than to create conditions with elevated RH and high air velocity. Elevated levels of RH favor growth of mold while high air velocity favors case hardening at the same time. Depending on air velocity inside the drying room, the amount of moisture removed from the surface of the product is vastly different. Fast moving air removes significantly more moisture from the surface of the product compared with slow moving air at the same level of RH.

If freshly filled salami is placed in the same drying room together with products, which have been dried or fermented for a few days already, the freshly filled product is hung in the upper area of the trolly and the products dried for a while already are placed near the bottom. Air traveling downwards along the walls inside the drying-room turns once it hits the floor and a slightly increased air velocity is seen near the floor as a result. If fresh products were placed in these lower areas, the risk of case hardening occurring is increased (Fig. 7.10).

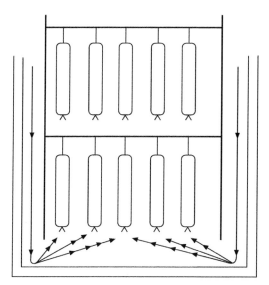

FIGURE 7.10 Fresh salami should be placed on the top layer while salami, fermented and dried for several days already, should be placed at the bottom due to higher air speed in the lower area.

Products, filled into different sized casings, present in the same room for fermentation or drying, commonly end up either with case hardening or mold on the outside. Case hardening in these cases occurs when moisture is removed too quickly from the surface of the large-diameter products but the level of moisture is too high for the small-diameter products, from, which moisture could be removed even faster. Generally, products with a similar particle size, filled into casings from similar diameter can be fermented at the same time. Of course, filling up a fermentation room with one and the same type of product would be perfect but this is not possible in many cases.

Even when the diameter of the casing used is the same, a significant difference in particle size of meat and fat particles present results in vastly different drying-behavior. Significant differences in particle size cause significant differences in behavior during fermentation and drying and products with different particle sizes can't be treated properly in the same room. Moisture has to be removed very slowly from super-fine salami due to the high density of the sausage mass as well as the large surface area of all the small-sized meat and fat particles. Slow removal of moisture from a coarse product, though, would most certainly cause growth of mold, and in this case moisture could be removed much faster from the surface of the product. Having a high Aw for a prolonged period of time within the coarse product means that *Lactobacillus* spp. has a long time in, which to ferment sugars and a lower pH value can be the result The loss in weight in the coarse product is also greatly slowed down and a much longer drying time is required to achieve the desired loss in weight. Products filled into larger-diameter casing generally have a lower final pH value compared with small-diameter products even though the same amount of acidification materials such as GDL or sugar, in combination with starter cultures, are added. This is because the Aw remains higher for a longer period of time in the core of the product and the lactic acid bacteria are more active producing elevated levels of gluconic acid. However, lower pH values do not automatically result in a firmer product and it is not the case that slightly more acidified products have a firmer texture. During drying, salami filled into larger-diameter casings lose less weight in percent within a given period of time than small-diameter product during. This is especially the case in the initial stages of fermentation as the surface area of larger-diameter products is smaller than that of smaller-diameter product. Small-diameter products demonstrate the largest surface area in relation to their weight.

Case hardening has to be avoided during fermentation and drying as it can lead to serious microbiological problems or other product shortcomings. It is not a problem from a microbial point of view if case hardening occurs in fast fermented salami after the pH declines to/below 5.2 within the first 36–48 hours as a pH of 5.2 or below stabilizes the product. Case hardening slows down the process of drying afterward, though. The desired firmness, or loss in weight, is in some cases never obtained or it can take considerably longer, which is a financial loss. Given that the pH value is generally at levels of 4.6–4.8 within fast fermented salami, and if case hardening occurs, heterofermentative *Lactobacillus* spp. will have sufficient moisture for a long time and large amounts of H_2O_2 and CO_2 will be produced. The development of rancidity is therefore speeded up and an enhanced number of pores, or tiny holes, can be seen in the product. In severe cases, the product can even burst due to large amounts of CO_2 produced. Obtaining case hardening in a later stage in the production fast-fermented salami is not a microbiological problem because all sugars have already been used up and the desired loss is weight has also been achieved. In fact, a slight degree of case hardening is occasionally produced purposely at the end of the drying process as sliceability of the product is improved as a result.

In medium-fast– and slow-fermented salami, if case hardening occurs in the early stages of fermentation and drying while the pH value, the main hurdle against microbiological spoilage, has not yet been established, the situation is different. The Aw in the product is too high for too long and this leads to serious microbiological problems because bacteria such as *Salmonella* spp., *Proteus* spp., *Citrobacter* spp., *Enterobacter* spp., *Escherichia* spp., *Pseudomas* spp., *S. aureus*, and especially proteolyic enzymes can grow for a prolonged period of time before the product has acidified sufficiently. Proteolytic enzymes produce metabolic byproducts, which are extremely alkaline and unwanted bacteria, in combination with prolonged enzyme activity, can dominate the total microflora. Therefore, the desired lactic-acid flora does not get a chance to produce sufficient lactic acid to stabilize the product from a microbiological point of view. In severe cases, the pH value never drops to desired levels and the product will be microbiologically spoiled presenting a serious health risk. On the other hand, if drying takes place too slowly, slime frequently forms on the surface of the product. This spoilage on the outside is mainly caused by excess levels of moisture in combination with high temperatures. Some slime producing bacteria are strongly proteolytic and the breakdown of protein leads to spoilage. Within the slime, bacteria such as *Staphylococcus* spp., *Micrococcus* spp., and yeasts are regularly present. If case hardening occurs in the later stage of drying when producing a medium-fast and slow fermented salami, this does not create a microbiological problem. The product is stabilized either by a pH value below 5.2 or by an Aw at or below 0.89. Case hardening always slows down further drying, though. The presence of a dry ring on salami, as is obtained when the case hardens, however, is not attractive to the consumer.

It must be emphasized that smearing of fat during processes such as cutting, mincing, mixing, and filling has to be avoided as the speed of drying is greatly reduced as a result. Once the speed of drying is reduced, the risk of obtaining case hardening is much greater. In summary, if case hardening occurs at the very beginning of fermentation stage, this can lead to microbiologically spoiled products especially in the manufacture of medium-fast– and slow-fermented salami. This is due to the high Aw persisting for a long period of time, which favors the growth of bacteria such as Enterobacteriacea and the entire fermentation process can never take place.

Drying of salami in larger factories can operate on a system in which climatic conditions in one fermentation or drying room are used to regulate climate in another drying or fermentation room. Several drying and fermentation rooms with different climates are in operation at the same time and salami in varying stages of fermentation and drying fill up these rooms. The benefit of this system is that when more moisture is required in one room, moisture can be drawn from another room with excess levels of moisture to the room requiring more moisture. The same applies to temperature and the exchangeable use of moisture and temperatures saves a lot of energy (and therefore cost) given that "creation" of a climate is always connected with the utilization of large amounts of energy. Processes such as increasing or decreasing moisture levels as well as increasing or decreasing temperature in large sized fermentation and drying rooms is very expensive and making use of the climate of other drying rooms reduces costs considerably. Of course, this system can only be used in companies, which have a large number of fermentation and drying rooms.

In large factories fermentation and drying rooms are in place where salami, or trollies filled with salami, is stacked in two layers. It has to be ensured that air flow is even in all areas within the room to avoid unwanted growth of would or case hardening (Figs. 7.11–7.15).

As described earlier a borderline between fermentation and drying is hard to be drawn because salami starts to lose weight already during fermentation as well. Generally, the start of drying is seen when fermentation, or any decline in pH value within the salami, has come to a stop. During drying, climatic parameters such as temperature, air speed and RH are gradually reduced and in the final stages of drying, the product is exposed to a temperature around 12–14°C, 70–75% RH and an air speed of around 0.05–0.1 m/s until the product reaches the desired weight loss in percent or the required Aw. Only then the product can be released for sale. The air speed must be greater than 0 m/s because unwanted

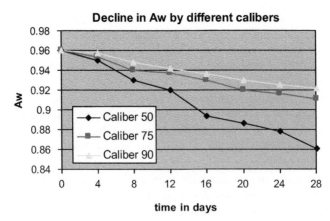

FIGURE 7.11 Impact of casing with different diameters on the reduction in Aw within salami.

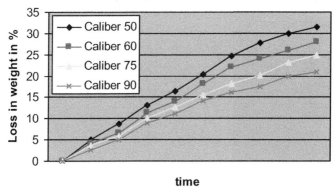

FIGURE 7.12 Loss in weight during drying by products filled into different-diameter casings.

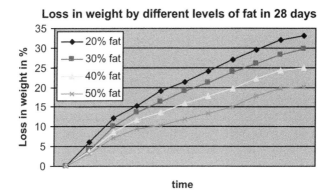

FIGURE 7.13 Loss in weight based on different levels of fat present within the product.

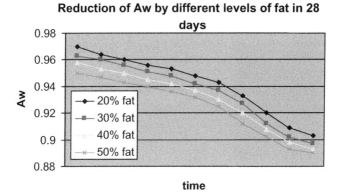

FIGURE 7.14 Reduction in Aw over time in salami based on different levels of fat within the product.

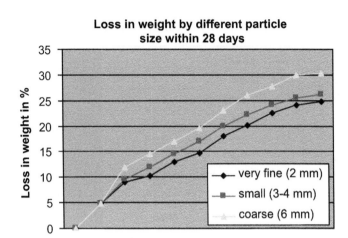

FIGURE 7.15 Impact of different particle size of meat and fat regarding the loss in weight during drying. (A) is a finely cut salami (2-mm particles size), (B) shows a particle size of 3–4 mm, and (C) exhibits a particle size of 6 mm.

mold would grow quickly if there was no air flow. Mold inhibiting substances in salami such as phenols and formaldehyde are volatile and their impact against growth of mold is reduced over time. Treating the product with materials such as sorbate or natamycin helps to inhibit mold growth during the final stages of drying, but mold growth can also be inhibited simply by having the proper climatic conditions in place. Drying is continuous until the desired loss in weight, or Aw, is achieved. In products stabilized against microbiological growth by its Aw value, the Aw reached has to be 0.89 or below (*Staphylococcus aureus* does not produce toxin at this level). In products stabilized by their pH value (5.2 and below) the Aw value is never important as a hurdle. It is important to dry the product to the desired Aw given that flavor, firmness, color and texture of the product are related to its Aw. On the other hand, excess drying is not beneficial in terms of the economics of production because less salami, based on weight, will be sold to the customer. The color of a dried salami is influenced by the loss of weight given that the material predominantly loses moisture during drying is muscle meat (and not fat). This causes nitrosomyochromogen within meat particles to be concentrated resulting in a darker color.

Recently a process called Quick-Dry-Slice (QDS) was developed and the principle behind this process is that individual slices of salami are dried to the desired Aw, or loss in weight, instead of drying an entire piece, or log, of salami in the first place to the point of being sliced into individual slices. This thought makes perfect sense as drying of slices is much more time-effective compared with the drying of a log of salami due to the significantly increased surface area of slices based on the same weight of salami to be dried. The process usually follows a fast-fermentation method reducing the pH value below 5.2 within 36–48 hours and the salami is most often "dried" for another 1–2 days before being placed by −1 to −3°C for several hours. The firm salami is peeled and sliced. Individual slices are taken up by robotic arms and placed on plastic trays. Trays enter a drying cabinet and slices are dried for a specified period of time to achieve the desired loss in weight. The tray with the dried slices is removed from the drying area and robotic arms place the slices into the packing machines. The entire process is continuous and the major advantage of this idea is that fermented and fully dried salami can be produced within 4 days instead of weeks. One downside is that the flavor of such salami is dominated by the acid-tangy taste from fermentation as well as the spices being applied and no typical buttery-ripened salami taste is seen. This is because there is no time provided for any enzyme activity to break down proteins or fat into free amino acids or free fatty acids.

7.3.7 Slicing

The majority of small-diameter raw fermented products such as salami snacks are not sliced and sold by piece instead. Products fermented in natural casings, and even large-diameter products, are commonly not sliced and prepacked but sold to deli shops or lunch counters where slicing takes place. Salami filled into medium and large-diameter fibrous casings are very often sliced and easy-peel casings (see Section 6.2 under Chapter 6: Casings) are used. Slicing takes place on high-speed slicing machines and a high degree of firmness, obtained through drying, supports sliceability. Commonly, the salami exhibits a slight degree of case hardening at the end of the drying period, which aids sliceability. Peeled and large-diameter products are stored at around 0°C for several hours before slicing to increase firmness of the product and sliceablity at the same time. Formation of condensation water should be avoided during slicing especially when dealing with products stabilized by an Aw below 0.89 and the pH value being above 5.2. If condensation water is present, then there is essentially 100% moisture on the surface of the salami and therefore there is sufficient free water for bacteria to grow in the outer layers. When condensation water forms during slicing of Aw-stabilized products, these products afterward have to be stored chilled at a temperature below +4°C to avoid bacteria growth. Condensation water is not such a big risk in products stabilized at a pH value below 5.2 as Aw has never been established as the main hurdle. The formation of condensation water is of no advantage, though, either.

7.3.8 Packaging/Storage

When sold in portions (rather than sliced) most salami products, stabilized either by Aw- or pH value, are commonly vacuum packed or packed under modified atmosphere. Some products sold in portions are occasionally packaged simply in nettings or in foil showing tiny holes, but further drying takes place when salami is packaged like this, which in most cases is not desirable. If these types of permeable packaging are used, they are applied to products, which are stabilized by an Aw below 0.89 as slow further drying reduces the Aw even further thus strengthening this hurdle. Sliced salami products, whether they are pH or Aw stabilized, are also vacuum or modified atmosphere packed (MAP packed). When vacuum packed, the individual salami slices tend to stick together and cannot be separated easily and therefore the individual slices end up badly deformed. Products of any diameter, small or large, which are stabilized via the impact of a low pH value (below 5.2) are predominantly packed under vacuum given that this type of salami has only lost between 15% and 20% in weight and no further loss in weight is desired. These products are commonly manufactured via fast and occasionally medium-fast fermentation and are generally products sold at a low price and excess drying is therefore not wanted. Salami, packed under vacuum, does not lose any additional weight after being packed and non-sliced, or portioned, products are generally sold based on a defined weight. Once the desired weight of the portioned products has been obtained during drying, vacuum packing takes place straight away. Any additional loss in weight is an economic loss given that the product, rather than a being a benefit in any way as the product, is already stabilized by its low pH value against microbiological spoilage. Salami products are also frequently flow wrapped and a type of bag is formed from a single-layered plastic sheet, which is finally heat sealed at both ends with the salami product being present inside the bag formed.

The packing materials used should have high oxygen and moisture barrier characteristics to avoid exchange in moisture between product and atmosphere. The permeability of packaging materials toward oxygen should be less than $10\,cm^3/m^2$ per day. Vacuum packing products inhibits growth of mold as mold requires oxygen for growth. Products that have been vacuum packed occasionally exhibit growth of mold once the packing is opened as internal moisture has been drawn toward the surface of the product during packaging and then the enhanced levels of moisture, especially on the surface, give mold an opportunity to grow. Also, products should not just be placed loosely in cartons, or boxes, by the box being closed afterward as some exchange in moisture between the salami and the air volume inside the box takes place. Because the box is closed no air velocity is taking place inside the box and moisture from the air inside the box is not removed. Mold will form on the surface of salami.

In MAP of salami, a mix of nitrogen as well as CO_2 is used. The level of CO_2 within the gas mixture is around 20–30% while nitrogen accounts for around 70–80%. CO_2 acts as a preservative in the form of carbonic acid and enhanced levels of CO_2 increase the acidity of the finished product. Carbonic acid is formed because CO_2 reacts with water within the salami forming carbonic acid ($CO_2 + H_2O \rightarrow H_2CO_3$). Oxygen should be present within an MAP pack at levels as low as possible and a level up to 1% is seen as the maximum. Any level of oxygen below 1% supports shelf life positively. Sliced salami packaged in a modified atmosphere lies loosely in the packaging and therefore it is not deformed by the packaging. Hence, individual slices do not stick together as it is the case of sliced products packed under vacuum.

Salami that is stabilized via a low Aw value (below 0.89) is commonly sold as a whole piece regardless of its size and diameter. It can also be sold in a sliced form, though. These products have lost around 30% or more in weight, are medium- to high-quality products, and have been fermented either slowly or at a medium-fast rate. A large proportion of salamis produced via medium-fast fermentation (during which acidification to a pH value of around 5.0 will have occurred) are subsequently been dried until around 30% in weight is lost thus reducing the Aw below 0.89. The combination of a pH value below 5.2 and an Aw below 0.89 makes these product double-safe. When the formation of condensation water has been avoided during slicing, products with a pH value of around 4.8–5.0, or an Aw below 0.89, are shelf stable without refrigeration. However, condensation water frequently forms during slicing and therefore most sliced products are stored and sold chilled at a temperature below +4°C.

A very critical point to remember is that the pH value of salami increases again during drying and storage. If salami is stabilized by a pH value between 4.9 and 5.0, the pH value increases over time as a result of proteolytic activity as highly alkaline metabolic by-products are produced. Once the pH value rises beyond 5.2, the low pH value being the main hurdle against microbiological spoilage, is lost and the product becomes microbiologically unsafe if the Aw has not dropped to a level of, or below, 0.89 at this point. Generally, maintaining a pH value below 5.2 is not a problem in fast- or medium-fast–fermented products. This is because a very low, and therefore safe, pH value of around 4.6–4.8 is usually seen in the finished product, and the pH remains well below 5.2 even during a prolonged period of storage. These products are also commonly consumed well before shelf-life expires or the pH value would rise beyond 5.2. Problems might occur in salami, in, which the drop in pH value only reached levels of around 5.0 as the pH value could rise beyond 5.2 within a relatively short period of time. At a pH value above 5.2, when an Aw below 0.89 has not yet been established, *S. aureus* is a serious microbiological risk as it will produce toxin. As a safeguard, salami stabilized via pH value must have a pH value below 5.2 at the end of the recommended shelf life or at point of consumption and not at point of sale bearing in mind that salami, sold from the manufacturer to the customer (supermarket) may remain on the shelves for weeks before being finally consumed. Salami stabilized via pH value are sold with a pH value around 4.6–4.8. This is still well below the critical point required for microbiological safety even if a slight increase in pH value occurs before the product is finally consumed.

Small white crystals are occasionally seen on the surface of salami and these crystals are made from creatine monohydrate. Crystal formation is found on the surface of products dried for a long period of time and creatine from meat itself is the source. The formation of white crystals happens commonly on the surface of packaged products, which are stored chilled, but crystals also form if the packed product is exposed to variations in storage temperature such as being stored in the chiller, placed for a while at ambient temperatures and then being placed again in the chiller. Creatine is very unstable in water and slight differences in moisture levels, intracellular or extracellular, cause creatine to be present either in its dissolved, or mineralized state. Mineralized creatine causes the white visible spots. One gram of creatine monohydrate requires 78 g of water to dissolve, ie, it has low solubility. These crystals are harmless for consumers but are not attractive and most consumers will not buy a product showing these crystals. So far, a full explanation of how to prevent the formation of these crystals, other than by avoiding variations of storage temperature, has not been produced.

7.4 SUMMARY OF CRITICAL PRODUCTION POINTS

- Meat and fat material processed has to show a low bacteria count of maximum 10^2–10^4/g and the pH value of the meat be 5.8 or below.
- No antibiotics should be present within meat and no hard MDM should be used.
- Fat must not be rancid and hard fat from loin or neck (if pork fat is processed), containing fewer unsaturated fatty acids, is preferred.
- Predrying of meat and fat reduces initial Aw.
- The level of added salt should be at least 26–28 g/kg of sausage mass.
- Nitrite or nitrite/nitrate to be added at the highest permitted level.
- Proper starter cultures in combination with sugars, or other acidifying material, should be added depending on the required speed of fermentation as well as the pH value reached after completing fermentation.
- Smearing of fat should be avoided during mincing, cutting, and mixing processes.
- The optimal temperature of the finished mass before filling is −4 to 0°C. Smaller particle–sized salami requires lower temperatures compared with coarser-sized products.
- Smearing has to be avoided during filling and casings are to be filled tightly applying medium-fast filling speed and full vacuum during the filling process. Casings have to be treated according to the manufacturer's recommendations and clips have to be attached firmly around the casing. The widest possible filling horn should be used in relation to the diameter of the casing to avoid back-curling of the sausage mass.
- Conditions during fermentation and drying vary between 93% and 72% RH, 28–12°C and air speeds of 0.8 to 0.1 m/s. The RH, temperature and air speed should gradually decrease during fermentation and drying to avoid case hardening.
- Cold smoke should be applied, if desired, for the first time after 36–48 hours once curing color is fully developed.

- Drying of small-diameter products containing coarse meat and fat particles can take place at a faster rate than the drying of products filled in large-diameter casings containing finely cut meat and fat particles.
- Salami is microbiologically shelf stable without refrigeration at a pH value below 5.2 or an Aw of, or below, 0.89.

7.4.1 Microbiology in Salami

It is almost impossible to completely summarize all the microflora within salami as raw fermented salami is by nature a "living" product and hundreds of millions of bacteria, from many different genera and species are involved. LAB are of utmost importance for acidification of salami. *Lactobacillus* spp. is the primary choice of bacteria used for acidification, which improves microbiological stability, color, taste, texture, and slice coherency of the product. Homofermentative *Lb.* are preferred to heterofermentative Lb. given that heterofermentative Lb. produce, besides some lactic acid, also acetic acid, H_2O_2 and CO_2. Some *Lactobacillus* spp. are blamed for the production of biogenic amines through decarboxylation of the amino acids tyrosine and histidine. *Pediococcus* spp., also a member of the LAB, also support acidification and both, *Lactobacillus* spp. as well as *Pediococcus* spp., seems to inhibit production of aflatoxins. Normally, salami have a Total Plate Count (TPC) of around 10^6/g at point of filling and lactic acid bacteria grow up to 10^8/g within the first 3–5 days of fermentation. Other lactic acid bacteria such as *Streptococcus* spp., *Enterococcus* spp. and *Leuconostoc* spp. are frequently present in salami at the beginning of fermentation. *Micrococcus* spp. and nonpathogenic species of *Staphylococcus* spp. have a major positive impact on the development and stability of the curing color as well as the flavor of salami.

Most of those bacteria produce the enzyme catalase, which slows down rancidity as it breaks down H_2O_2 into water and oxygen. *Micrococcus* spp. in particular reduces nitrate very effectively to nitrite thus helping formation of a strong curing color. *Micrococcus* spp. and *Staphylococcus* spp. are found commonly on the surface of dried products as they require oxygen for living. *St. aureus* has to be tightly controlled by pH and/or Aw as it can cause serious microbiological problems. An Aw at/below 0.89 or a pH at/below 5.2 controls *St. aureus* and inhibits toxin production.

Salmonella spp. are regularly present in meat and are controlled by an Aw below 0.95 or a pH value below 5.5. In fast- and medium-fast–fermented salami, an Aw below 0.95 is obtained within a few days of fermentation and therefore *Salmonella* spp. is inhibited at an early stage. *Salmonella* spp. presents a greater risk in slow-fermented salami as hurdles, such as Aw and pH value, are put in place at a later stage of fermentation. *Campylobacter* spp. is very sensitive to reduced Aw levels and is generally not a risk in salami. *Clostridium botulinum* is a risk in raw fermented sausage mainly in products such as summer sausage, which is fermented at temperatures above 30°C. When poorly cleaned natural casings are used for the production of salami, *Clostridia* spp. can cause problems as they grow underneath the casings within the product. *Leuconostoc* spp. can form viscous strings when sucrose (dextrane) is present in combination with high fermentation temperatures (26–30°C). *Listeria monocytogenes* is regularly present in meat as well as in salami but during fermentation and drying the number should not exceed 10^2/g of product. However, some countries have food standards in place, which demand *L. monocytogenes* to be negative in 25 g. Generally, raw fermented salami does not support the growth of *L. monocytogenes*, enterohaemorrhagic *E. coli* (EHEC) and *Salmonella* spp. and these bacteria are even reduced during fermentation as a result of acidification to pH values at/below 5.0. *Listeria monocytogenes* is inhibited in salami by a pH value below 5.0 as wells as the presence of a strong competitive flora mainly consisting of *Lactobacillus* spp.

Some molds produce the enzyme cellulose. This enzyme, predominantly produced by *Mucor* spp., can break down cellulose in fibrous and other casings. In severe cases the enzyme literally "eats" the casing. *Mucor* spp. can be prevented or inhibited by the application of a proper fermentation and drying climate (not too humid) and the application of smoke at the right time before any mold is seen on the product. Treatment of the salami with sorbate or natamycin also prevents growth of mold. Final drying at an RH below 75% reduces the risk of mold growing given that mold-inhibiting substances commonly seen on salami such as phenols and formaldehyde are volatile and lose their effectiveness over time.

Yeasts are present within the salami mass at the beginning of production in low numbers and can only grow on the outside layers of salami since they require oxygen for growth. Proper control of the fermentation climate and the application of smoke eliminate unwanted yeast.

Chapter 8

Typical Fermented Non-Heat Treated Salami Products Made Around the World

8.1 HUNGARIAN SALAMI

Hungarian salami is generally made from pork meat. The meat processed is predominantly from heavy sows that are older than 2 years, weighing around 180–200 kg. This type of meat is chosen because it contains less free water, is darker in color, and is firmer than meat from young pigs of around 6 months of age. In general, starter cultures were not used in the past to make this kind of salami, but they are commonly introduced nowadays. When a starter culture is applied, a small amount of sugar, mostly dextrose, is added as well to ensure proper color and flavor development and to guarantee some degree of acidification. Very few other additives are introduced, most commonly only nitrite and spices such as pepper, nutmeg, and cloves. The meat is generally cut in the bowl-cutter until the particles are 2–3 mm in size and then the additives are mixed with the cut meat.

Fermentation starts at low temperatures, for example 10–12°C, and the temperature is kept low until an Aw below 0.95 is reached in the product, which makes it stable against *Salmonella* spp. The pH value drops as a result of natural acidification to around 5.2, and from this point pH is a hurdle against *Staphylococcus aureus*. Having the combination of a pH of 5.2 and an Aw below 0.95 ensures that the product is safe from a microbiological point of view. The salami is smoked with natural smoke at low temperatures for a short period during the initial stages of fermentation and dried.

After around 1½–2 weeks, mold is sometimes applied to its surface. Often, though, placing the salami into rooms with other molded salami is sufficient to cause the internal house flora to grow on its surface. The presence of mold and a long period of drying result in the typical buttery flavor, which is due to compounds formed through proteolysis as well as lipolysis. In the finished product, which is commonly 90–100 days old, the level of fat is around 45% and salt is present at around 4%. The peroxide number varies between 0.6 and 0.9 and the TBA value (see Section 1.9 under Chapter 1: Meat and Fat) is between 0.8 and 1.0. Given that proteolytic enzymes have been active over a long period of time, the pH of the finished products is generally between 6.0 and 6.3. An Aw of 0.85–0.88 is also achieved, and therefore the product is shelf stable.

8.2 KANTWURST (AUSTRIA)

Kantwurst is a rectangular type of raw fermented salami made in Austria. The pH value of Kantwurst declines to around 4.8–5.0 within 3–4 days, and therefore it is a medium-fast fermented product. Acidification occurs due to the introduction of starter cultures, which ferment dextrose into lactic acid. Commonly, the raw meat and fat materials are cut in the bowl-cutter to a particle size of around 2–3 mm. Pork (sow meat) is predominantly used, and up to 20% beef is occasionally processed as well. Additives introduced are salt (26–28 g/kg), nitrite, ascorbate, and spices such as black pepper, garlic, coriander, and caraway seed. Dextrose is the sugar of choice, and the sausage mass should be at a temperature of between −3 and −1°C once it has been removed from the bowl-cutter. It is subsequently filled into fibrous casings with a diameter commonly of between 55 and 75 mm. The product is not tightly filled; rather it is only filled to around 80% of total capacity.

The filled sausage is then placed layer-by layer into a press. The press is divided into layers by partitions made of stainless steel that are around 3–4 cm in height along their longer side. The salami are usually as long as the partitions are wide. The relatively loosely filled salami is placed in layers in the press tightly next to each other so that they do not move or stretch while they are pressed. The partitions are used to separate the layers: once a layer is full, another dividing layer of stainless steel is placed on top of it, which is then filled up with salami. Eventually, the entire press is filled with many layers of salami, each layer containing several individual pieces, and once the press is full, a thick layer of stainless steel is placed on top of the final layer. Gentle but firm pressure is then applied to the salami by tightening the press' two screws or by using compressed air. Finally, the press is placed in the fermentation room.

During the first 2–3 days of fermentation, the screws are tightened several times so that the pressure applied to the product is continuous. During this period of time, the temperature in the fermentation room is between 20 and 24°C and the relative humidity (RH) is around 90%. After around 4–5 days, during which time the pH has fallen to around 4.8–5.0, the product becomes sliceable and is removed from the press. Slime is commonly seen on the product as moisture cannot escape from the product while it is trapped in the press. The product is washed with a weak and lukewarm salt solution (3–4%) and hung.

At this stage, the salami already is rectangular in shape as a result of being pressed. During the 4–5 days of fermentation, the sol inside the product has also been transformed into a gel via acidification. The product is hung in a room at a temperature of around 20°C, an RH between 86% and 90%, and under an air-speed of around 0.6–0.4 m/s. As soon as the surface is dry, the product is cold-smoked at 20–25°C to avoid growth of mold. Smoking is repeated several times over the following 2–3 days, and the product is dried until around 25–30% in weight is lost. During drying the rectangular shape become even more pronounced. The finished product is stable and has a pH of around 5.0–5.2 and an Aw of around 0.90. Products can be dried for a longer period, in which case they have an Aw value of around 0.88–0.89 (30% loss during drying). They are therefore shelf stable without refrigeration.

8.3 GERMAN SALAMI

In a typical German salami, pork is present in around 70–80% of the total recipe with beef accounting for around 20–30%. The salami exhibits a particle size between 2 and 3 mm, and most often a bowl-cutter is used to achieve this degree of granulation. In large-scale operations, the mincer-filling head system can be seen often. The pH value of German salami declines to around 4.8–5.1 within 3–4 days, and therefore it is a medium-fast fermented product. Acidification takes place due to the introduction of starter cultures. Additives introduced are salt (26–28 g/kg), nitrite, ascorbate, and spices such as black and white pepper, garlic, coriander, and caraway seed. The sausage mass should exhibit a temperature of between −3 and −1°C once it has been removed from the bowl-cutter abd efore it is filled under vacuum into fibrous casings with a diameter commonly of between 85 and 95 mm.

After the conditioning period, fermentation commences by 22–24°C and the RH is around 90–92%. Give that this product is filled into a large-diameter casing and exhibits a small particle size, the removal of moisture has to take slowly in order to avoid case hardening. After around 3–4 days, during which time the pH has fallen to around 4.8–5.1, the product becomes sliceable. Cold smoke at a temperature around 25°C is applied several times over the next couple of days in order to obtain a strong smoke color and smoke flavor. German salami does not display noble mold, and the strong impact from the smoking process supports the absence of mold.

Temperature and RH are gradually reduced over the next 3–5 days, and final drying of the salami occurs at a temperature of 12–15°C, an RH between 72% and 75%, and an air velocity of 0.05–0.1 m/s. German salami loses around 30–34% in weight during drying resulting in an Aw-value of around 0.88–0.89. They are therefore shelf stable without refrigeration.

8.4 LUP CHEONG (CHINA)

Lup cheong has been produced in China for thousands of years. It used to be made of goat and lamb meat mixed with onions, salt, and pepper. Today, though, the product is made of coarsely minced pork meat, pork fat (or fatty pork meat), sugar (up to 10%), salt (2.2–2.8%), Chinese wine, soy sauce, five-spice mix (fennel, watchau, anise, cinnamon, clove), nitrite, and MSG. The main flavor comes from soy sauce and the high levels of sugar, which make the sausages taste sweet. A sweet taste is desired as Asians generally do not favor an acid or a sour taste in meat products.

To produce Lup cheong, fatty pork trimmings are usually minced with the 8- to 10-mm blade, and all the additives are mixed with the minced trimmings. Water is added at around 10–15% to support the formation of wrinkles during drying in the finished product. The mixed mass is filled into 24- to 26-mm casings and dried above charcoal for 24–48 hours at 45–55°C and a low RH (around 70%). Such intensive drying, in conjunction with high levels of Aw-reducing additives such as sugar and salt, quickly lowers the Aw to 0.90 within 2–3 days. The pH always remains high as lactic acid bacteria do not play a significant role in the process: their numbers are only around 10^3–10^4/g of sausage. After the initial fast reduction in Aw, the product is left for 3–4 days for further drying and equilibration at room temperature. Another result of such fast drying is that the sausage has the desired wrinkly appearance and the characteristic reddish color with large white particles of fat. Occasionally, the product is dried even further and the Aw in the finished product can be as low as 0.75–0.8. The pH of the finished product is generally between 5.7 and 6.0.

Lup cheong is predominantly sliced diagonally, fried, and consumed hot with vegetables or rice. It is microbiotically stable due to the combination of high levels of salt and sugar and the rapid reduction in Aw (Aw in dried product is below 0.90), made possible by filling the sausage into small-diameter casings and drying under high temperatures.

8.5 CACCIATORE (ITALY)

Cacciatore is made of 100% pork meat, and predried meat from sows is often processed. It is also common to work with fatty pork meat trimmings rather than sow meat and sow meat can also be mixed with pork trimmings in order to achieve a high-quality product. The fat content in the unfermented mass is around 25%, and additives such as salt (2.8%), nitrite, ascorbate, and sugar are added. Spices such as black and white pepper, coriander, garlic, and some chili are frequently used. Starter cultures are also introduced, and the raw meat together with dextrose and fat materials processed should be slightly frozen. The meat and fat materials are minced with the 8- to 10-mm plate, and all is mixed in a paddle-mixer until the mixture is slightly tacky. Proper distribution of all additives, cultures, and salt has to be ensured during mixing. The sausage mass exhibiting a temperature between −2 and 0°C is subsequently filled under vacuum into beef or pork runners with a diameter of 35–37 mm and hand-tied with a string. The length of an individual piece is around 12–15 cm. The formed products is dipped into a solution containing noble mold if mold is desired on the final product. Fermentation takes place for around 2 days at a temperature of 24–26°C and an RH of 90–92% until a pH between 4.8 and 5.0 is reached. Application of mold via spraying is also frequently practiced after around 3–4 days postfermentation. Following that, the temperature is reduced to around 18–20°C and the RH lowered to 84–88%. Final drying and storage take place at 12–15°C and around 75% RH until about 30–32% is lost in weight. The product is shelf stable at the end of the process at an Aw below 0.89.

8.6 SOPRESSA (ITALY)

Pork meat only is processed when producing Sopressa and sow meat is the material of choice. The fat content is around 20–25% in the nonfermented meat mass resulting in a fat level of between 30% and 35% in the finished product after drying is completed. The meat and fat materials are either minced with the 10-mm plate and placed in a mixer before all additives are applied or the mincer-filler head system is applied instead. Additives chosen are salt at 26–28 g/kg of meat mass, starter cultures, dextrose, nitrite, and ascorbate, as well as spices such as cracked black pepper, garlic, coriander, and a small amount of chili. When working with a mixer, the salami mass is mixed with all the additives, spices, and starter cultures until a slightly tacky salami mass is obtained. Salt and nitrite are added last during the mixing process, and even distribution of all materials has to be ensured. Once mixing is completed, the salami shows a temperature between −2 and 0°C. The salami mass is filled under vacuum into large fibrous casings with a diameter of 110–120 mm or beef bungs. The casing is not filled tight and only to around 80% of its capacity. The underfilled casing is placed in cages that are spring-loaded in order to apply pressure from two sides. As a result of such pressure, an oval-shaped salami is obtained.

Fermentation starts at 24–26°C and an RH of 90–92% for 2–3 days during which the pH value drops to 4.8–5.0. The temperature and RH are then gradually lowered to 18–20°C and 80–85%, respectively. Final drying takes place at 12–15°C, an RH of around 70–74%, and an air velocity of 0.05–0.1 m/s. Noble mold is frequently applied to the product by mold growing on the surface of the salami as a result of the house flora present. However, in most cases mold is applied by dipping the salami in a mold-containing solution after the product was filled into the respective casing. Another method of applying mold is to spray the salami after around 3–4 days of fermentation. During drying, Sopressa salami experiences a loss in weight of 30–34% and the product is shelf stable as an Aw below 0.89 is achieved.

8.7 MILANO SALAMI (ITALY)

Milano salami is made of 70–75% lean pork (90–95 CL-grade) from the shoulder and 25–30% pork back fat. Salt is added at 2.8–3.0%, sugars at 5–7 g/kg and spices at around 5 g/kg. The spices most commonly used are white pepper, cardamom, and garlic. Red wine is frequently added as well at around 1 L/100 kg of sausage mass. The meat to be processed should be semifrozen and the fat should be frozen. The meat and fat materials are cut in the bowl cutter to a particle size of 3–4 mm and should be at a temperature of −3 to −1°C once the sausage mass is obtained. The sausage mass is then filled under vacuum into fibrous casings of 80-mm diameter or natural casings. After filling, a netting is put over the salami and handmade nets are used for top quality products. The fermentation process starts with tempering for around 2–5 hours at 22–24°C and 60–65% RH in order to remove condensation water. Then the product is fermented at 22–26°C and 90–94% RH for 2–3 days. The temperature and RH are then gradually lowered to 18–20°C and 80–85%, respectively. Final storage and drying take place at 12–15°C, an RH of around 70–75%, and an air velocity of 0.05–0.1 m/s. Mold is applied to the product. This can take place straight after the product has been filled, in which case the product is dipped into a mold-containing solution. More often, though, the product is smoked slightly (using cold smoke at 20–25°C) and then mold is sprayed on 2–3 days later. The total loss in weight during drying is 30–34% and the product is shelf stable as an Aw below 0.89 is achieved. Milano salami only acidifies very slightly: the pH drops during fermentation to a level of around 5.2–5.3.

8.8 SUCUK (TURKEY)

Sucuk is a ring- or horseshoe-shaped delicacy heavily consumed in Turkey. It is an air-dried fermented product made of mixture of meat and fat material originating from beef, mutton, (sheep) and buffalo. In the past, though, sucuk was made of beef only. Besides spices and salt, no other additives are introduced in products produced in the traditional way. Salt is added at around 25 g/kg and nitrite is applied as well. The main spices are cumin, black pepper, fresh garlic, ginger, cinnamon, and cloves, and sugar is added as well. Meat and fat materials are coarsely ground, and all additives and spices are added before being well mixed. Commonly, the sausage mass is left standing in the chiller overnight before being filled into beef or sheep casing of 24–28 mm. The filled product is then smoked several times before being dried for 2–3 weeks. Nowadays, 0.4–0.5% of GDL are commonly added to the product to speed up fermentation and to increase shelf-life of the product. Often, a strong garlic and cumin flavor, rather than the typical fermentation flavor, is seen in the final product. Some degree of rancidity in the flavor is also accepted by the consumer. Sucuk is cut into slices, fried on both sides, and primarily consumed with scrambled eggs for breakfast.

8.9 CHORIZO (SPAIN)

There are countless different types of chorizo produced in Spain, and almost every region has its own version. Chorizo can be made of pork only but a mix of beef and pork is also very common. The fat content of the freshly made product is around 30% and spices such as paprika (which gives the red color to the product) is used at fairly high levels. Other spices such as chili (hot or less hot), garlic, and pepper as well as salt (2.5%), nitrite, and ascorbate are part of the product. The meat and fat materials are cut to around 6–8 mm in diameter before the sausage mass is filled into pork- or beef runners. Occasionally, the sausage mass is also filled into 50- to 70-mm fibrous casings. The product is slightly smoked after fermentation and subsequently dried to varying degrees. Some products are heavily dried until an Aw-below 0.89 is obtained. Others are less thoroughly dried. Chorizo is consumed raw, fried, or cooked.

8.10 FUET (SPAIN)

Fuet is a Catalan product from Spain and generally made only of pork. The fat content is around 30–35% and the particles of meat and fat display a size of 3–5 mm. Spices utilized such as paprika and garlic give the product a sweet and aromatic flavor. The sausage mass is filled predominantly into collagen casings of 38–46 mm in diameter and occasionally into natural casings as well. Noble mold is applied to the surface and the fermented product is dried until around 30–35% in weight are lost.

8.11 PEPPERONI (USA)

Pepperoni in the United States is a raw sausage made of beef and pork or pork only. Products made of 100% beef must be called beef pepperoni. The loss in weight during manufacture of pepperoni is around 30%, ending up with a water:protein ratio <1.6:1, and fermentation takes place primarily under the impact of fast-acidifying starter cultures and citric acid is also frequently applied to conform with the desired speed of acidification. The fat content of pepperoni is around 30–35%. To manufacture this type of salami, semifrozen meat and fat materials are cut in the bowl-cutter to a particle size of 4–5 mm before salt and nitrite are added. The final granulation is 2–3 mm. Spices, starter cultures, color enhancers, and other additives are introduced into the bowl-cutter at the beginning of the cutting process. Paprika is commonly added to give a touch of red within the product. Salt is present at around 27–29 g/kg of sausage mass. The tacky mass is filled into easy-peel fibrous casings of 42–47 mm in diameter. Back-rolling has to be avoided during filling as this can lead to cupping of the sliced product on the pizza (see Section 7.3.2 under Chapter 7: Fermented Salami: Non–Heat Treated). The widest filling pipe in relation to the diameter of the casing has to be used to avoid back-rolling of the sausage mass.

Fermentation starts at 32–37°C and 90–92% RH for around 24 hours and continues at 90% RH and a temperature of around 20°C for another 24 hours. On the third day, the temperature is reduced to around 18°C and the RH lowered to around 85–87%. Cold smoke is occasionally applied during the second and third days in order to obtain some degree of some color and smoke flavor. Drying continues by 13–16°C, an RH between 70% and 75%, and low air velocity until a water:protein ratio of 1.6:1 is obtained.

8.12 PIZZA SALAMI (AUSTRALIA)

In Australia, starter cultures have to be applied by law when producing raw fermented salami and the pH value has to decline below 5.2 within 48 hours after fermentation started. Pizza salami is generally produced of fatty pork trimmings and meat is minced with the 6- to 8-mm plate; however, a small amount of pizza salami is also made using the bowl-cutter and the particle size is 2–3 mm. The fat content of the final product is around 30%. To manufacture this type of salami using minced meat and fat materials, semifrozen meat and fat materials are minced and all minced materials are placed in a mixing machine before spices, starter cultures, dextrose, ascorbate, salt, and nitrite are added. Paprika is commonly added adding a touch of red/orange color to the product. Salt is present at around 26–28 g/kg of sausage mass and all is mixed until some degree of tackiness is seen. The tacky mass is filled under vacuum into easy-peel fibrous casings of 40–42 mm in diameter. Back-rolling has to be avoided during filling as this can lead to cupping of the sliced product on the pizza (see Section 7.3.2 under Chapter: 7 Fermented Salami: Non–Heat Treated). The widest filling pipe in relation to the diameter of the casing has to be used to avoid back-rolling of the sausage mass.

Fermentation takes place at around 26–28°C and 90–92% RH for around 48 hours and continues at 90% RH and a temperature of around 20°C for another 24 hours. On the fourth day the temperature is reduced to around 18°C and the RH lowered to around 82–85%. Cold smoke is applied for 2–6 hours after 48 hours mainly to suppress the growth of unwanted mold. Temperature and RH are reduced gradually and final drying occurs by 12–15°C, an RH between 72% and 75%, and an air velocity of 0.1 m/s until the total weight loss is around 30%.

8.13 BEEF SALAMI (MALAYSIA)

Fermented salami is not widely produced in Asia because Asian people usually dislike a sour-acidic taste in meat products. To produce this type of product, the material of choice most often is boneless beef brisket with the fat on. The particle size of the meat and fat in the end product is between 2 and 3 mm, and the salami is produced either by using the bowl-cutter or the mincer-filling-head system. Acidification occurs mostly due to the introduction GDL, and starter cultures are normally not applied. Additives introduced are salt (25–27 g/kg), nitrite, ascorbate, and spices such as black pepper, garlic, and coriander. The sausage mass should be at a temperature of between −3 and −1°C once it has been removed from the bowl-cutter or when processed over the mincer-filling head. It is subsequently filled under vacuum into fibrous casings with a diameter commonly of between 55 and 65 mm.

Fermentation starts between 26 and 28°C and the RH is around 90%. After 36–48 hours, the pH value has dropped to around 4.8–5.0 and a small degree of cold smoke is applied. During the next 4–6 days, the temperature is gradually reduced to 12–15°C, the RH is dropped to 70–74%, and air velocity is 0.05–0.1 m/s during the final stage of drying. Total weight loss is between 25% and 30% and the end-product is generally vacuum packed as well as stored under chilled conditions when offered for sale.

Chapter 9

Fermented Salami: Semicooked and Fully Cooked

9.1 MANUFACTURING TECHNOLOGY

Given that time is always associated with cost, and this is especially the case when products such as salami are dried and fermented, semicooked and fully cooked products are a good medium between raw fermented (non–heat treated) and fully cooked products. Semicooked and fully cooked fermented salami are predominantly produced following the fast- or medium-fast fermentation process (see Sections 7.3.3.8 and 7.3.3.9 under Chapter 7: Fermented Salami: Non–Heat Treated). The sausage mass is obtained using either a bowl-cutter or a mincer-mixing system. In finely cut products, frozen and semifrozen meat and fat materials are most commonly cut in the bowl-cutter to a granulation size of 2–4 mm. The fat content of the product is around 30–35% and pork is the type of meat most often processed. However, other types of meat such as beef and lamb, or a mix of pork and beef, are processed frequently as well. Additives such as glucono-delta-lactone (GDL) (8–10 g/kg), salt (around 25 g/kg) nitrite, ascorbate, and spices are introduced. Salt and nitrite are added last during mixing to obtain a small amount of sol. Even though the product is eventually partly or fully cooked, starter cultures are also commonly introduced into the sausage mass, in combination with sugars, to obtain a pH value of around 4.8 within 48 hours of acidification. The temperature of the finished sausage mass is around −3 to 0°C. Smearing of fat has to be avoided during cutting or mincing. The sausage mass is subsequently most often filled into natural or fibrous casings of 42- to 70-mm diameter, with fibrous casings being the most common choice. A long resting time prior to filling has to be avoided in case GDL is applied as GDL hydrolyzes into gluconic acid, especially on the top layers of the nonfilled salami mass placed in trollies or tubs. Smearing has to be avoided during filling because the appearance of the salami is improved if meat and fat particles can be distinguished in the final product as a result of nonsmearing.

Fermentation of the filled product starts by conditioning for around 1–5 hours at 22–24°C and an RH of 60–70% to remove condensation water. Following conditioning, RH is increased to around 90–92% for the next 24–28 hours to support formation of gluconic acid from GDL or the activity of starter cultures. Temperature is increased as well to 26–28°C to speed up formation of gluconic acid. In countries such as the United States, encapsulated citric acid is applied and acidification takes place during heat treatment. If starter cultures are applied in the United States, it is common to work with very fast acting bacteria, reducing the pH value within the salami to levels around 4.8 within 24–36 hours and temperatures around 35–38°C are applied during fermentation. After 36–48 hours, the pH value has dropped to levels around 4.8 and sol has been denatured by the impact of acidification and turned into a gel by a pH value of 5.2. Therefore, the product is sliceable. Nitrosomyoglobin has been denatured and stabilized as well once the pH reaches 5.2. Cold smoking of the fermented product is frequently the next step, but not necessarily if no smoke flavor is desired, and the product is smoked at 20–25°C several times over the next 24–48 hours until the desired smoke color is obtained.

The smoked product is then thermally treated with either steam or dry heat (baking) at 70–75°C until temperatures of around 55–60°C are reached in the core of the product for a semicooked product. Food laws vary from country to country and, for example, if an acidified salami in Australia is cooked to 65°C in the core and this core temperature is held for 10 min, the product is considered to be fully cooked. Application of dry heat is preferred as it results in strong color in the final product while removing large amounts of moisture at the same time. When temperatures of 55–60°C are obtained in the core, the outer areas of the product are almost fully cooked while the core remains semicooked and even partially raw. The thermally treated product is not dried any further; however, in some countries, drying continues at 12–14°C and 72–75% RH for another 3–5 days. A loss in weight of around 15–20% is commonly the target before such semicooked or fully cooked products are released for salt. Large-volume and price-sensitive products from this product group are commonly produced within 1 week, covering all processing steps such as cutting/mixing, fermenting, heat treatment, and final drying.

The finished product is occasionally sold as a whole piece, but most often slicing is the last processing step before being offered for sale. The product is stored under refrigeration at temperatures below +4°C even though it has a pH value below 5.2, thus being stabile from a microbiological point of view. Following such a procedure results in a semicooked or fully cooked product that still exhibits some degree of salami character and flavor to a large degree but has been produced in a short period of time. However, several countries do not permit the production of these semicooked/semiraw products reaching 50–55°C in the core, insisting instead that a product is either raw fermented and dried (non–heat treated) or fully heat treated.

Chapter 10

Typical Fermented Semicooked and Fully Cooked Salami Products Made Around the World

10.1 SUMMER SAUSAGE (USA)

Summer sausage is produced from lean beef, pork, pork fat, or fatty pork trimmings. The particles of meat and fat in the finished products have a diameter of around 3–4 mm, and the fat content of the product is around 30%. Additives such as salt (around 25–28 g/kg), nitrite, and spices are added, and black and white pepper, mustard seed, nutmeg, garlic, coriander, and allspice are the spices most frequently used. Nitrite and fast-acting starter cultures are also introduced. Frequently, citric acid is also applied to conform with the required drop in pH value. The sausage mass is filled into fibrous casings of 70–80 mm diameter, and the product is fermented at high temperatures of 30–40°C. The pH drops quickly as a result, and the final pH obtained is usually around 4.5, which is low but accepted by the customer. After being smoked, the salami is heated to around 60–65°C in the core at low RH levels (around 40–50%). The product is dried further until the desired weight loss has occurred.

10.2 PIZZA SALAMI (AUSTRALIA)

Fermented and heat-treated pizza salami is Australia is generally produced out of a mixture of lean and fatty pork or fatty pork trimmings only. The particles of meat and fat in the finished products have a diameter of around 2–3 mm and the fat content of the product is around 30%. Comminution of meat and fat takes place using the bowl-cutter or the mincer-filling head option. Additives such as salt (around 26–28 g/kg), nitrite, and spices are added, and black pepper, nutmeg, garlic, and paprika provide the desired red-orange color to the end product. Dextrose is the sugar of choice and is added at around 6–8 g/kg of salami. Nitrite, ascorbate, and fast-acting starter cultures are also introduced. The sausage mass is filled into fibrous casings of 40–60 mm diameter, and the product is fermented at temperatures of 26–28°C and an RH of around 90–92%. The pH value drops below 5.2 within 48 hours, and the final pH obtained is usually around 4.5. The fermented salami is not smoked but dried for another 4–7 days before being heat treated at around 70°C until a core temperature of

65°C is obtained. Maintaining the 65°C in the core for 10 min results in a fully cooked end product. The product is showered for a few minutes before being placed in the chiller. Once chilled to a temperature below +5°C, the product is usually peeled, sliced, and packed. The finished product is stored under chilled conditions at temperatures below +4°C.

10.3 POULTRY SALAMI (ASIA)

Within Asia poultry, salami is produced using boneless chicken or turkey meat from thighs with the skin on. The meat is processed when still semifrozen, and mincing of the meat materials to the desired particle size is most often practiced before being mixed with all spices and additives. Occasionally, the mincer-filling head system is applied and glucono-delta-lactone is generally applied to achieve acidification to a pH of around 4.7–4.9. Salt is applied at around 24–26 g/kg, and the salami is mildly spiced. MSG is introduced as flavor enhancer, and nitrite is also applied. Ascorbate acts as the color enhancer, and the filled salami is fermented at around 24–28°C under an RH of around 90% for 2° days. Within those 2 days, the pH drops to 4.7–4.9, and smoke is applied at 55–60°C for around 1 hour before the salami is cooked to 70°C in the core, applying 75°C as the cooking temperature. Most often, the application of steam is the choice of cooking method. The product is cooled down and stored under chilled conditions at temperatures below +4°C.

Chapter 11

Nonfermented Salami: Fully Cooked

Nonfermented salami (NFS), which is cooked strictly speaking, has nothing in common with salami because "salami" is by definition a non–heat-treated product. Unlike salami, NFS products do not acidify and water is even added during the manufacturing process, rather than being removed through drying. NFSs can be shelf stable without refrigeration but are often stored at temperatures at or below +4°C. Large or small visible particles of lean meat as well as small particles of fat present the show meat, because those particles are seen in the end product. Such show meat in NFS is held together most often by a base -emulsion (BE). However, for simplicity reasons NFS is also produce without a BE in order to eliminate the additional process of obtaining a BE. Finished products can be categorized into finely cut/minced or coarse-minced products, and all products are filled into nonwaterproof casings such as cellulose, fibrous, or textile casings.

11.1 SELECTION AND PREPARATION OF RAW MATERIALS

NFS are predominantly made of pork. They are occasionally made of beef and lamb, though, for consumers who do not consume pork for religious reasons. Pork trimmings are processed for show meat, and the meat to be processed must have a low bacteria count; a number between 10^2 and 10^4/g is optimum. The fat, or fatty trimmings, to be used must not be rancid and their microbiological status should be comparable to that of the lean meat. It is preferable to use hard fat containing a low number of unsaturated fatty acids (such as fat from the loin, neck, or fatty bellies), as the risk of smearing during cutting or mincing is increased if soft fat from shoulder or leg is used and clearly cut particles of fat should be visible in the finished product.

The 95 chemically lean (CL) grade lean beef is commonly used for the production of a fine BE, which acts as the glue within the product between the individual pieces of meat and fat. Using dark-firm-dry (DFD) beef to make the BE should be avoided if possible, as this type of meat can affect color development negatively, but if some of the beef processed is of DFD character there will be no significant disadvantage. The amount of BE within the total product is generally between 20% and 30%. In an all-pork product, BE can be made from lean pork meat. It is not of significant advantage to have some PSE-character pork as the visible particles of meat. However, the concentration of PSE pork within the total amount of pork used should not exceed 10–15%. Commonly meat from cuts such as shoulder, neck, and rump are used and these cuts are very dark, thus hiding a small amount of PSE pork. In high-quality NFS, meat from the leg is processed as well. If this type of meat is used, once again, sow meat is the preferred choice. Boar meat should not be used as must people are very sensitive to the smell of boar, caused by the male hormone androsterone, and find it unpleasant.

All meat materials to be processed must be stored at temperatures below +4°C to avoid or delay growth of bacteria such as *Staphylococcus aureus*, *Pseudomonas*, and *Salmonella*. The meat and fat material to be processed for finely cut products are stored in the freezer. Just as in the production of raw fermented salami, fat is primarily processed fully frozen, whereas meat is processed semifrozen. The temperature of the fat to be processed for finely cut products should be around −18°C, whereas the meat should be at a temperature of around −5°C. Pork fat from loin and neck is also processed chilled as it is hard fat containing a low amount of unsaturated fatty acids. Such chilled fat is precured and stored in the chiller at a temperature below +4°C; precuring hardens the fat even more.

11.2 SELECTION OF ADDITIVES

NFSs contain very few additives as substances such as carrageenan, soy protein, colors, starch, or any other filler (binder) are not introduced. The only additives usually applied are phosphates, salt, nitrite, color enhancer, and spices. Salt is the most basic additive and is applied at levels of between 18 and 20 g/kg of product. The main reasons for adding salt to NFS are its contribution to flavor and its contribution to activating protein in the manufacture of the BE. Salt at those levels is not a hurdle against microbiological spoilage as it is in raw fermented products. Nitrite is added to ensure a stable strong curing color develops and for its contribution to curing flavor. The level of nitrite added has to be adjusted according to the maximum level permitted in the finished product in respective countries' food standards. Around 60% of the nitrite added to the uncooked product is used for the development of the curing color and flavor, and a fair amount is also lost during heat treatment. Some nitrite is oxidized to nitrate during processing and is therefore not available after thermal treatment to stabilize curing color (the enzyme responsible for the reduction of nitrate to nitrite [nitrate reductase] is destroyed by temperatures above 68°C).

Color enhancers are commonly added, ascorbate or ascorbic acid being the ones usually chosen. The level of ascorbic acid/ascorbate is normally around 0.5–0.6 g/kg of total mass. Care has to be taken that color enhancers based on ascorbic acid do not come in direct contact with nitrite as highly toxic nitrose gas (NO_2) would form, and nitrite would not be present afterward for the formation of curing color. Nitrose gas is extremely toxic and presents a serious health risk.

Phosphates are applied at 4–5 g/kg of BE, and the BE is the finely cut tacky mass formed acting as an effective binding glue between the meat and fat particles. In case cooked salami is produced without BE, phosphates are introduced at 3–5 g/kg of total salami. The type of phosphates applied is cutting phosphates, activating a high amount of muscular protein during the short period of cutting in the bowl-cutter. These contain a large proportion of short-chained phosphates, which activate protein rapidly. Spices and herbs are introduced according to the manufacturer's recipe. Garlic, pepper, and coriander are often the main spices introduced. In some products, visible spices such as cracked black pepper or green peppercorn are used for contribution to the appearance of the final product.

11.3 MANUFACTURING TECHNOLOGY

To produce NFS, particles of fat and lean meat of various sizes are mixed together, and several different manufacturing methods are commonly practiced. In products with large lean pieces of meat as show meat, precuring is recommended as the time for color development in the finished product is then significantly shortened.

11.3.1 Precuring of Meat and Fat Materials

To precure the lean meat (which is generally pork of 95 CL grade but can also be lean beef or poultry meat in nonpork products) or meat trimmings of any type, all the necessary nitrite and salt are first added. Around 18–20 g of salt and around 120–200 ppm of nitrite are applied per 1 kg of meat (depending on the respective food law in place). The particles of meat to be precured are around 20–40 mm in diameter. The meat and additives are mixed well to ensure even distribution of salt and nitrite, and the mixture is left standing in the chiller for development of the curing color. When fatty pork belly is used as the visible meat, the belly meat is cut up into smaller pieces and precured in the same way. The advantage of precuring is that once the material is actually used for the production of NFS, all color development in larger pieces of meat has already taken place. The introduction of salt also causes fiber structures in the lean meat to swell. Therefore, water present in the muscle tissue does not leak out and the precured lean meat suffers no weight loss. Another benefit of precuring is that the introduction of salt and nitrite slows down bacterial growth on and in the meat.

The fat used in minced products is commonly precured as well but not for color development as no myoglobin is present within fat. The addition of salt and nitrite slows down bacteria growth on the fat and therefore increases its shelf-life under chilled conditions. The addition of salt to materials such as lean meat and fat also increases its firmness, thus reducing the risk of fat smearing during mincing. Cleanly cut particles of lean meat are therefore present in the finished product. The downside of precuring is the additional processing step requiring time as well as storage of the precured meat. This is the reason why large companies opt not to precure and rather having a longer heat-treatment process to ensure full development of curing color in the end product.

11.3.2 Manufacture of BE

The BE (*Grundbrat* in German) is the binding glue between the particles of fat and lean meat in the finished product. A BE is generally produced from lean beef or pork (95 CL grade), water/ice, and additives such as salt, nitrite, and phosphates. In most cases, the level of extension is around 30% and, for example, 30 kg of water/ice is added to 70 kg of lean beef/pork to obtain 100 kg of BE. For 100 kg of BE, around 18–20 g of nitrite (180–200 ppm/kg), 400–500 g of cutting phosphate, and 2 kg of salt are added. The process starts by placing the lean meat in the bowl-cutter and cutting it under medium-fast knife speed at around 1500–2000 rpm. Phosphates, around 70% of the total ice and water, and the nitrite and salt are added. The relative amounts of ice and water added at this stage are dependent on the temperature of the lean meat. A temperature of 0°C is desired before the cutter can be switched to a fast knife-speed at around 3000–3500 rpm. In case chilled meat is processed, basically all of the 70% is ice. Some water can be included, though, when the beef processed is slightly frozen. All the materials are cut under high knife speed until the temperature is around +4 to +6°C. At this point, the remaining 30% ice is added.

Ice is added rather than water at this point to reduce the temperature of the BE within the cutter to around 0°C. All the materials are then cut under high knife speed by around 3500 rpm until they are at a temperature between +2 and +4°C, to complete the process. At the end, a highly tacky, shiny, and finely cut emulsion is obtained. The temperature range during the process is purposely kept between 0 and +4°C as activation of protein is most effective at these temperatures and bacterial growth is controlled as well. The BE can be stored in the chiller at temperatures below +4°C for several days as long as the mass was at a temperature below +4°C when it was removed from the bowl-cutter.

11.3.3 Finely Cut Cooked Salami

A product with small particles of fat and meat is obtained commonly using one of two possible methods. One method is to cut the frozen fat and semifrozen meat material in the cutter, while the other is to obtain the desired particle size by mincing the meat and fat materials. The particle size of fat and meat in this type of product is most often between 2 and 3 mm but can be as small as 1 mm.

A finely cut salami is produced in a way similar to a raw fermented salami, at least in the initial stages. A typical recipe contains 25–30% BE, 25% pork back fat, and around 45–50% lean meat, most commonly pork. Frozen fat is placed in the bowl-cutter and cut with sharp knives for a short period until particles of around 8–10 mm are obtained. Semifrozen lean meat is added to the fat, and all the materials are cut under medium-fast knife speed at around 1500 rpm until the desired granulation is almost reached. As in the manufacture of finely cut salami, it is important for the frozen materials to flow freely while being cut. Therefore, the bowl-cutter should only be around 50% full so that materials are not pushed around or squeezed during cutting. Once the desired granulation size of the meat and fat is obtained, the speed of knives is reduced to around 200–300 rpm and the BE is evenly added into the cutter. All the spices, color enhancer, and salt (and nitrite) are also introduced at the level required for the meat and fat materials, taking into account that the BE already contains salt and nitrite (spices and color enhancers are added at a level appropriate for the total amount of sausage as the BE does not contain these already).

All is mixed well to ensure even distribution of all spices and additives until a tacky meat mass is obtained. The application of vacuum during mixing is also advantageous. The vast majority of air pockets are removed by mixing under vacuum, and a shiny as well as compact sausage mass is obtained. Removing oxygen also helps in the development of the curing color in the final product. The final temperature of the sausage mass prior to filling is around 0 to +2°C, which prevents smearing of fat during filling. It can be beneficial to cut the frozen fat and semifrozen meat as the finished cooked product then looks more like a salami than if the mass is minced. On the other hand, not all fat and meat particles in the finished product are exactly of the same size if the material is cut, as they are if the sausage mass is minced.

Another method of obtaining a small-particle NFS is to place the BE into the bowl-cutter and add precured fat and meat materials to the BE. All the materials are cut under slow speed, and spices and color enhancer are added. The level of spices and color enhancer is calculated according to the total batch size. No further salt or nitrite is added because the BE, as well as the precured meat and fat materials, contains salt and nitrite already. Commonly, though, meat and fat materials are processed to be present in the end product as show meat that are not precured and therefore salt and nitrite has to be added at a level to cover all of the show meat. Adding precured show meat has the advantage that curing color is already developed within the meat and the process of heat treatment is shortened. Also, precured meat is significantly firmer than meat and fat materials, which are not precured, supporting a clean cut during the mincing process.

The entire mass is cut for a few turns in the bowl-cutter to ensure even distribution of the fat, meat, and all additives. Then the sausage mass is removed from the cutter and minced with the 2- to 4-mm plate using a double set of knives. The knives and plates inserted into the mincer are placed in the following sequence: precutter ⇒ knife ⇒ 10- to 13-mm plate ⇒ knife ⇒ desired final plate (2–4 mm) ⇒ fixation ring. Care has to be taken that all fittings are tightened up properly and a clean cut of all the meat and fat materials is obtained without smearing of fat. The speed of mincing is moderate so that the sausage mass passes comfortably through the mincer without being pushed. Commonly, the minced sausage mass is placed afterward either into a paddle-mixer or into the bowl-cutter to be mixed for a short while under vacuum. After mixing, the final temperature of the sausage mass is between +4 and +8°C but can be lower if frozen fat is used to support a clean cut. The advantage of mincing the entire sausage mass is that every single fat and meat particle is of the same size in the finished product.

Another "simple" method of obtaining an NFS with a particle size between 3 and 10 mm is to place all meat and fat materials, which are not precured, into a bowl-cutter and cut at a knife speed of around 700–1000 rpm until particles of 15–20 mm are obtained. During cutting of the meat and fat materials, all additives such as phosphates, salt, nitrite, color enhancer, as well as spices are evenly added. The level of all additives and spices is based on the total batch size. On top of that, around 7–9% of chilled water is added during the cutting process. The addition of phosphate and salt supports the uptake of the water by water, in turn supporting the activation of protein. Once a granulation size of 15–20 mm is obtained, and all additives and spices are within the meat mass, all is mixed gently for around 2–3 minutes until a tacky meat mass is obtained. No further cutting of meat and fat should take place during the mixing process. The tacky meat mass is removed from the bowl-cutter and minced with the desired plate using a double knife set (as explained earlier). The minced meat mass is then mixed either in the bowl-cutter or in a paddle-mixer to ensure even distribution of all meat and fat particles as well as increasing the degree of tackiness.

In both methods, in order to support the mincing of a tacky meat mass, some cold water is sprinkled on the mincer, which makes the meat mass "slip" better toward the mincer worm. Also, once all materials are minced, it is of benefit to add couple of kilograms of raw or precured meat into the mincer, which "clears out" all remaining salami mass stuck between the plates and knives within the mincer.

When producing a small particle size minced salami, it is advised to use a double set of knives because reducing the size of meat and fat particles from, for example, 15–20 mm to 2–3 mm in one mincing step increases the risk of fat-smearing, resulting in an "unclear" smeary definition between meat and fat in the end product. In severe cases, the smearing of fat can lead to fat separation during the heat-treatment process.

11.3.4 Coarse Minced-Mixed Cooked Salami

The production of coarse minced NFS can be achieved in several different ways. One of them is to place the BE in the bowl-cutter and add frozen fat. Cutting commences under medium-fast knife speed at around 1500 rpm, and salt and nitrite are added at an appropriate level for the amount of fat present (unless the meat to be added later is not precured, in which case the level of salt and nitrite needs to be higher). Spices are introduced at a level appropriate for the total batch size. The fat is then cut to the desired granulation, which is commonly around 3–5 mm. Afterward, minced lean meat (8- to 20-mm blade) and predominantly precured, is gently mixed in, sometimes under the impact of vacuum.

Large amounts of coarse NFS are manufactured using another method. This involves placing the BE in the bowl-cutter and adding chilled, frozen, or even precured chilled fat. All the ingredients are cut for a short while until minceable pieces of fat are obtained. Additives such as spices, color enhancer, and nitrite are added to this mass. The BE/fat mass is then removed from the bowl-cutter and commonly minced with the 5- to 8-mm plate at a moderate speed using sharp mincer knives to obtain a clean cut without smearing. Once mincing is completed, the mass is placed into a paddle-mixer and precured lean meat, minced with the 10- to 20-mm plate, is added. All the materials are mixed gently for a short while possibly under a vacuum. This process ensures, first, that all particles of fat are of the same particle size in the finished product and, second, that the particles of fat are smaller than the particles of lean meat, which makes a good contrast in the final product.

By following a recipe containing only around 20% BE, 20% fat, and 60% lean meat, a very attractive and high-quality product is obtained. Visual spices such as cracked pepper or green peppercorn are regularly mixed into coarse NFS as well.

11.3.5 Filling

The sausage mass obtained is filled subsequently into permeable casings. Fibrous casings of a diameter between 60 and 120 mm are predominantly used. Casings made out of linen, other textiles, collagen, or cellulose are also used in high-quality products, and a small percentage of products are filled into large-diameter natural casings, such as beef bung. Fibrous casings are generally soaked prior to use in lukewarm water for around 30–45 minutes and any other type of casing utilized should be treated according to the manufacturer's recommendations. Filling takes place under a vacuum to eliminate all residual air trapped in the sausage mass. This ensures that an air-free product with good slice coherency is obtained. The application of a vacuum during filling also supports the formation of a strong curing color. Filling speed is largely determined by the caliber (diameter) and type of casing used as well as the particle size of meat and fat. Generally, larger-caliber casings are filled at a slower speed than small-diameter casings, but the filling speed in general is moderate in order to avoid smearing of fat. As mentioned, the speed of filling has to be adjusted so that smearing is avoided and fat and meat particles are clearly visible in the filled product. The casings are filled tightly and clipped, or tightly closed with a string. Care has to be taken that the clip does not cut the casing as it is secured around the casing. Even the tiniest of cuts made to the casing cause it to burst during heat treatment. Clips, attached too tightly, can also cut the casing, and as a result, the hung sausage will drop during thermal treatment.

11.3.6 Drying, Smoking, and Cooking

Once filling is completed, reddening is the first step in thermal treatment in case any nonprecured meat material is processed. The reddening step allows for the development of curing color and takes place by 50–55°C and an RH around 80–85% and lasts between 30 minutes and 1 hour depending on the diameter of casing. At 50–55°C, no protein is denatured, fully supporting the development of the curing color. The next processing step is drying, which takes place at around 60–65°C and low RH (around 40%) for approximately 30 minutes to 1 hour. The drying time has to be adjusted according to the caliber of the casing, and generally, products filled into a large-diameter casing require longer drying. The filling level of the smoking chamber has an impact on the speed of drying. Moisture is removed at a slower rate from each individual pieces of product when the chamber is full compared to when the chamber is only half full. Temperatures between 60 and 65°C also speed up the development of curing color tremendously, and once the surface of the casing is dry and a strong red curing color is seen, smoking can commence. The period of drying is extended when nonprecured lean meat is used as show meat, especially if the meat was minced with a 13- to 20-mm blade. This is because when non precured meat is used, the drying process is the period of time in which curing color develops. Products filled into a large-diameter casing and containing large pieces of nonprecured meat (13–20mm) require a drying period of up to 2 hours to ensure full and proper color development in the core of the product. To avoid such prolonged periods of drying, though, precured lean meat is frequently used as the show meat. Whatever the type of meat used, curing color has to be fully developed before smoke is applied.

Smoking is the next step, and the temperature applied is between 65 and 75°C by an RH of around 50–70%. Smoking continues until the desired smoke color is seen on the product. Generally, it lasts for 1–2 hours, but strongly smoked products, filled in large-diameter casings, can be smoked for up to 3 hours. Upon completion of smoking, the product is either fully cooked with steam or baked with dry heat. Final thermal treatment with steam, applying temperatures of 76–80°C so that a core temperature of 72°C is reached, completes the process and the cooked product is showered for a few minutes to avoid formation of wrinkles during cooling. NFS can also be treated with dry heat at 78–80°C to reach temperatures up to 72°C in the core; this is another method that is frequently practiced. Cooking the product with steam results in a paler color on the surface than does cooking with dry heat, because steam washes off some of the smoke applied to the surface during smoking. Applying low levels of moisture during all processing steps such as drying, smoking, and cooking, on the other hand, results in a lovely color as well as a strongly flavored product. Products thermally treated with dry heat at the end of the process are either showered for a few minutes to avoid the formation of wrinkles or left unshowered when a natural-looking wrinkly product is desired.

Often, finished products that were treated with dry heat during "cooking" are dried for another 7–21 days at 10–12°C and an RH around 72–74% if the respective food law in place allows it. This is done so they lose more weight despite the fact that the application of dry heat during the cooking period has removed large amounts of moisture. However, in many countries, a nonacidified and fully cooked product has to be stored at chilled conditions at temperatures below +4°C. Drying of the product after thermal treatment, if permitted, reduces the Aw to around 0.92, which makes the product shelf stable without refrigeration because vegetative bacteria belonging to the group of Enterobacteriaceae such as *Salmonella* are killed by temperatures of 72°C and would require an Aw above 0.95 to survive. *Staphylococcus aureus* is also generally killed at temperatures of 72°C, but surviving bacteria could form toxins at an Aw around 0.92 and as such would need an Aw below 0.9 from a food safety point of view if stored under ambient temperature. Spore-forming bacteria such as *Clostridia* also require an Aw above 0.95 and as such are also deactivated because the Aw is at around 0.92.

11.3.7 Packaging and Storage

The type of packaging and especially storage temperatures used vary significantly, depending mainly on whether the product is nondried, partially or fully dried, and sold as a whole piece or sliced. Products sold as whole pieces that have not been dried any further after thermal treatment (nondried products) are generally vacuum packed and stored below +4°C, as their Aw is not safe below 0.95. Vacuum packing prevents the growth of aerobic spoilage bacteria as well as mold and avoids any further loss in weight. The product is then generally sliced in the shop prior to being sold. Nonsliced products that have been dried further after thermal treatment have an Aw between 0.92 and 0.93 and are commonly vacuum packed. These are also commonly stored refrigerated even though there is no need as the Aw below 0.95 stops the growth of bacteria such as *Salmonella*. Some nonsliced products are dried for such a long period of time that the Aw drops to a level of, or below, 0.89, which makes the product microbiologically stable from an Aw point of view, because bacteria such as *S. aureus* do not produce toxin at Aw values at or below 0.89. These fully dried, nonsliced products are also commonly vacuum packed, and as long as no condensation water forms during packing, they can be stored at room temperature. Products with an Aw of 0.92 or even below 0.90 are frequently stored under chilled at around +4°C just so there is another hurdle in place against possible bacteria growth. Care also has to be taken that the packaging foil used has high moisture-barrier characteristics.

The situation is different if the product is sliced before being packed. Nondried sliced products are always stored at temperatures below +4°C to delay, or inhibit, growth of any bacteria introduced on to the product during the slicing process. Slicing of semidried or fully dried salami carries the risk of the formation of condensation water. This is especially the case if the product to be sliced was placed in the freezer or in a tempering room at temperatures around −2 to 0°C to increase sliceability of the product before slicing. The slicing-room has a significantly higher temperature than the cooled product and therefore condensation water is likely to form when the product is brought out of the tempering room or freezer. Condensation water on the surface of salami represents an Aw of 1.00 and spores as well as bacteria can grow as a result once they have access to condensation water regardless of the internal Aw of the product. As soon as condensation water forms, Aw is no longer present as a hurdle and sliced products must be stored at temperatures below +4°C to avoid bacteria growth. Semidried products with an Aw around 0.92 and especially fully dried products with an Aw of below 0.90 are firm in texture. They are occasionally sliced without being cooled beforehand (ie, they are taken straight from the drying room to the slicing room). If this is the case, formation of condensation water is avoided, but the sliced products are still commonly stored under chilling conditions to avoid growth from bacteria as bacteria may have been introduced during the slicing process itself.

11.3.8 Summary of Critical Production Points

- Meat and fat material used to show low bacteria count, between 10^2 and 10^4/g.
- Fat processed not to be rancid and hard fat, low in unsaturated fatty acid, is preferred.
- BE to have a maximum temperature of +5°C.
- Sharp bowl-cutter knives are essential when producing cut products. The fat in minced products must not smear during the mincing process.
- Precuring of meat materials used for show meat is recommended.
- The product should be tightly filled into nonwaterproof casings under a vacuum and subsequently dried at around 60–65°C, smoked at 65–75°C, and thermally treated to a core temperature of 72°C predominantly by applying dry heat at 76–80°C.
- Products thermally treated with dry heat are dried until an Aw of 0.92–0.93 is obtained, which makes the products shelf stable without refrigeration as all vegetative pathogens able to survive at an Aw below 0.92 have been destroyed during heat treatment. Products that are dried until an Aw of, or below, 0.89 is reached are fully shelf stable by the low Aw. Formation of condensation water during packing of the product to be avoided in products that are not stored refrigerated.
- Steam-treated and nondried products are vacuum packed and stored refrigerated at a temperature below +4°C.
- Most sliced products, regardless of internal Aw, are vacuum packed and stored at a temperature below +4°C as the product may have been contaminated during slicing.

Chapter 12

Typical Nonfermented Salami Products Made Around the World

12.1 POLISH SALAMI (AUSTRIA)

Polish salami in Austria contains around 25% base emulsion (BE), 25% fat, or 35% fatty pork belly, and around 50% lean precured pork. The BE is generally cut together with the fat, or fatty pork belly, in the bowl-cutter for a few turns, and spices such as garlic, pepper, and coriander are added. Salt and nitrite are introduced into the BE fat material at a suitable level to cover the nonprecured fat material (the salt content of the finished product should be around 2%). The BE including the fat, or fatty material, is minced with the 4- to 6-mm blade, while the precured lean pork is minced separately with the 13-mm blade. Both minced materials are then mixed together in paddle-mixers and commonly filled into 75- to 90-mm fibrous casings. Visible spices such a black cracked pepper or green peppercorns are introduced during mixing as well. Green peppercorn in-brine has to be washed and drained before being introduced into the sausage mass. This is because the peppercorn brine is acidic and would interfere with binding, causing the added peppercorns not to be effectively bound. Other spices introduced are garlic, ground black pepper, coriander, and nutmeg. The filled product is then dried at 65°C for around 1 hour, smoked at 70–75°C for 1–1½ hour, and, finally, cooked with dry heat at 80°C until 70–72°C is reached in the core. After a short shower, the product is placed in the chiller. The majority of Polish salami is stored afterward at a temperature below +4°C as it is not shelf stable. Some producers dry the thermally treated product at a temperature of 12–15°C and an RH of 72–75% until an Aw of around 0.92 is obtained, which makes the product shelf stable without refrigeration if packed as a whole piece avoiding formation of condensation water. Polish salami is also occasionally steam cooked after smoking. If that is the case, the steamed products are always stored refrigerated as further drying normally does not take place.

12.2 CHEESE SALAMI (AUSTRIA)

A cheese salami is more or less the same as Polish salami except that 15–20% of cubed cheese is added to the sausage mass. The type of cheese regularly applied is one with a high melting point such as Swiss Emmental. The cheese is cut into cubes of 1 × 1 cm before being mixed into the mass. The cheese chosen must not melt during thermal heat treatment at around 70–72°C.

12.3 VIENNA SALAMI (AUSTRIA)

Vienna salami contains 25% BE as well as 75% precured fat and meat materials that are mixed together in a way that the fat content in the end product is around 20%. Precured meat and fat material of pork are cut in the bowl-cutter together with the BE for a few rounds under medium-fast knife speed while all additives such as color enhancer and spices are added. The salt content of the salami is around 1.8–2.0%, and the main spices are ground white pepper, garlic, nutmeg, coriander, and a hint of glove. The mixed and coarse-cut meat mass is removed from the bowl-cutter and minced with the 3-mm plate. A clean cut has to be obtained during mincing and smearing of fat is to be avoided. The minced sausage mass is commonly then gently mixed afterward under a vacuum before being filled into fibrous casings of 75- to 90-mm diameter. Heat treatment is very similar to Polish salami as described under point 1.

12.4 SALAMI FLORENTINE (AUSTRIA)

To make this type of cooked salami, the first processing step is basically the same as when producing raw fermented salami, and the total batch consists typically of 20% BE and 80% meat and fat with the fat content being around 25% in the final product. Frozen fat and semifrozen meat are cut in the bowl-cutter using sharp cutter knives until a granulation of 2 mm is obtained. The bowl-cutter is only around 50–60% full (50–60 kg of meat and fat in a 100-L bowl) at this point as the frozen and semifrozen materials have to flow freely without being squeezed. BE as well as all the required additives and spices are added into the cut meat and fat particles, and all is mixed well under slow knife speed until a tacky mass is obtained. As the meat and fat materials processed for show meat are not precured, salt and nitrite are added for the amount of meat and fat materials processed. Color enhancer is also introduced at a level appropriate for the total mass including BE. The level of salt in the finished product is around 1.9–2.0%. The well-mixed meat mass is filled into fibrous casing with a diameter of 90–110 mm and heat treatment is very similar to that for Polish salami as described under point 1. Care has to be taken that the drying step is around 1.5 hour to ensure full development of curing color because the show meat (around 80%) is not precured and the meat mass is very cold at the point of filling.

12.5 CABANOSSI (AUSTRIA)

Cabana in Austria is a high-quality product made of 25% BE and 25% fat. The remaining 50% is precured lean meat of 90–95 CL grade. Sometimes, pork bellies of 75 CL grade are used, and these contain all the fat and lean meat material required. Most commonly, all materials such as BE, fat, and lean meat, as well as fatty pork bellies, are mixed in the bowl-cutter for a few turns, and additives, salt, and spices are introduced. Cabanossi enjoys a rustic mixture of black pepper, lots of garlic, and some chili. The mass is removed from the cutter, minced with the 4- to 6-mm blade, and most commonly filled into 24- to 30-mm collagen casings.

Filled sausages vary in length but are mostly between 20 and 25 cm. The first step in their thermal treatment is drying for 15–20 min at low RH (around 40–50%) and 60–65°C. Smoking the takes place at 65–70°C for 20–30 min followed by the application of dry heat at around 80°C until a temperature of 72°C is obtained in the core. Once cooled, the cooked products are dried at 12–14°C and 72–74% RH until 30% in weight are lost. They are then vacuum-packed and are shelf stable without refrigeration. However, despite being shelf-stable, they are most often stored under chilled conditions.

12.6 STRASBURG SALAMI (AUSTRALIA)

Strasburg salami is Australia is a price-competitive product, and a typical recipe would contain 20% chicken MDM (mechanically deboned meat), 20% water, 10% reworked material from other products, and around 50% pork trimming of 75 CL grade. All meat and fat materials are usually minced with a coarse plate and all, as well as water, are mixed in a paddle-mixer. Additives such as phosphates (3 g/kg of total mass), salt (16–18 g/kg of total mass), nitrite, color enhancer, spices, soy protein isolate (around 1–1.5%), as well as starch (2–3%) are introduced during mixing. The spice applied provides a gentle flavor based on white pepper, mace, nutmeg, a hint of garlic, and some cardamom. All is mixed until a cohesive mass is obtained before passing the mixed meat mass through an emulsifier having inserted a 6- to 8-mm plate as the desired particle size. The emulsified meat mass is filled under vacuum into waterproof casings with a diameter of 65–80 mm. The product is finally steam-cooked at 80°C until 70°C is reached in the core.

12.7 CHICKEN SALAMI (ASIA)

Boneless chicken thigh meat with skin-on is the material most often processed and is minced with the 8- to 10-mm plate before being placed in the paddle-mixer or on the bowl-cutter for mixing purposes. During the mixing process, all additives, very similar to Strasburg salami, are added. The exception is that soy protein is not applied because the salami is low in fat and only contains around 10% added water. The mixed meat mass is minced with the 3- to 5-mm plate using a double set of knives to ensure a clean cut of all meat and fat materials during the mincing process. Postmincing, the meat mass is mixed again in the paddle-mixer or bowl-cutter to ensure even distribution of the fat and skin particles. Filling takes place under vacuum, and fibrous casings with a diameter of 65–80 mm are processed. Heat treatment is very similar to that of Polish salami (see point 1) with the only difference that the final cooking step takes place by applying steam at 80°C, and not dry heat, in order to reach 72°C in the core. This type of salami is always stored under chilled conditions because the Aw never represents a hurdle against microbiological spoilage.

Index

'*Note*: Page numbers followed by "f" indicate figures, "t" indicate tables.'

A

Abbatoir, 39
Acetyl coenzyme A (Acetyl-CoA), 31
Acid carboxyl group, 6–7
Acidification, 147, 178–179, 184
   additives selection, 123
      chemical acidulants, 123–124
      sugars for acidification and flavor, 124–125
Acidity of meat, 49–50
Actin, 10–11, 13, 14f, 18, 31
Activated acetic acid, 32–33
Additives, 59, 185, 187, 201
   antimold materials, 70–71
   antioxidants, 78–80
   caseinate, 65–66
   chemical nonbacterial acidulants, 86–87
   colors, 84–86
   emulsifiers, 59
   fibers, 87–88
   flavor enhancers, 72
   HVPs, 78
   liquid smoke, 84
   MSG, 71–72
   natural smoke, 80–83
   phosphates, 59–62
   pork rind powder, 68
   potassium chloride, 64–65
   ribonucleotide, 72
   salt–sodium chloride, 62–64
   sodium-reduced sea salts, 65
   soy protein, 66–68
   spices and extracts, 74–78
   sugars, 68–70
   water/ice, 72–73
   whey protein, 65–66
Additives selection, 119–120, 190–191
   acidification, 123
      chemical acidulants, 123–124
      sugars for acidification and flavor, 124–125
   ascorbic acid, 121–122
   case hardening, 122–123
   nitrate, 121
   nitrite, 120–121
   selected wanted mold and yeast, 130–132
   spices, 122
   starter cultures, 125–130
Adenosine diphosphate (ADP), 16–18, 31–36, 38, 40
   levels postslaughter and rigor mortis, 38
   differences in meat preslaughter and postslaughter, 38t
   post mortem glycolysis, 38
   rebuilding of ATP steps, 31
      glycolysis, 32–33
      krebs cycle, 33–35
      oxidative phosphorylation, 35–36
ADI. *See* Available daily intake (ADI)
ADP. *See* Adenosine diphosphate (ADP)
Aerobic bacteria, 91–93
Air velocity, 141–142, 146
ALA. *See* Alpha-linolenic acid (ALA)
Alapyridaine, 72
Albumins, 11
Aldose, 68, 69f
Alkaline, 6
Allium cepa. *See* Onion (*Allium cepa*)
Allium sativum. *See* Garlic (*Allium sativum*)
Alpha-carbon atom (Cα), 4
Alpha-carbon atom. *See* Chiral center
Alpha-linolenic acid (ALA), 24–25, 25f
Alternating program, 152
Amadori rearrangement, 54–55
Amalgamating flavors, 75
Amino acids, 4, 6
   alpha-carbon atom, 4
   buffer capacity, 6
   configuration, 5f
   essential, 5
   food, nutritional value of, 5–6
   L-or D-form of amino acid alanine, 5f
   peptide bond formation, 7f

203

Index

Ammonium ion (NH_4^+), 33
Anaerobic glycolysis, 36–37
   lactic acid formation, 37–38
   level of glycogen, 37
   post mortem glycolysis, 37
Anaerobic process, 32
5α-Androsten-16-en-3-one, 112–113
Androstenone. *See* 5α-androsten-16-en-3-one
Angkak, 85–86
Animal psychiatrists, 39
Antimold materials, 70–71
Antioxidants, 78–79, 121
   ascorbate, 79
   heavy metal ions, 80
   phenolic substances, 79
   polyphenols in green tea oil, 80
   sage, 80
   tocopherols, 79–80
Argon, 39
Aroma, acidification impact on, 147–148
Artificial casings, 107. *See also* Natural casings
   fibrous casings, 107–108
   flat width, 108
   permeability of fibrous casings, 108
   woven textile casings, 108
Ascorbate, 79, 100–101, 121–122, 188
Ascorbic acid, 78–79, 98, 100, 121–122
Atomization, 84
Autoxidation, 27
Available daily intake (ADI), 59
Aw value, 51

B

Baader machine, 43–44
Bacterial growth, 191–192
Bacterial proteolysis, 150–151
Bacteriostatic properties of garlic and thyme, 74
Bag of spice, 74
Base–emulsion (BE), 189
    manufacturing, 192
BE. *See* Base–emulsion (BE)
Beef casings, 103–104
Beef fat, 117–118
Beef salami, 184
3,4-Benzopyrene, 81
Betanin, 85
BHA. *See* Butylated hydroxianisole (BHA)
BHT. *See* Butylated hydroxitoluene (BHT)
Biochemistry of meat. *See also* Meat biochemical processes
    in meat postslaughter, 36–38
    in meat preslaughter, 31–36

Black pepper, 76
Bladders, 103–105
Boar meat, 112–113
Bowl-cutter, salami in, 133
   CO_2, 135–136
   methods of cutting salami, 134
   salami with mincer-filling head, 137–138
   salami with mincer-mixing system, 136–137
   sol protein, 134
   temperature, 135
Brine method, 152
Bronchothrix, 91
Bungs, 103–104
Butylated hydroxianisole (BHA), 79
Butylated hydroxitoluene (BHT), 79

C

Cabanossi (Austria), 200–201
Cacciatore, 180
Candita famata (*C. famata*), 130
Cap, 105
Capillary effect, 15, 87–88
Capsaicin, 75
Carbon dioxide (CO_2), 39, 91, 126
Carbon monoxide (CO), 92
α-Carbon, 22–23
Carbonyls, 82–83
Carboxy-myoglobin, 92
Care, 139
Carmine, 85
Carminic acid, 85
Carnosic acid, 80
Carnosol, 80
Carotenoids, 76
Case hardening, 159, 166–167
Caseinate, 65–66, 159
Casings, 103
   artificial casings, 107–108
   natural casings, 103–107
Catechine, 80
Cattle, 41–42
Cell, 47
Chain–phosphates, 60
Cheese salami, 199–200
Chemical acidulants, 123–124
Chemical nonbacterial acidulants, 86
   encapsulated citric acid, 87
   GDL, 86, 86f
Chemically lean (CL), 189–190
Chemiosmosis, 35
Chicken MDM. *See* Mechanically deboned meat from chicken (Chicken MDM)

Chicken salami, 201–202
Chili, 75
Chilled meat, 113
Chinese restaurant syndrome, 72
Chiral amino acids, 4
Chiral center. *See* Alpha-carbon atom (Cα)
Chloride ions, 65
Cholesterol, 22
Chorizo (Spain), 182
Citric acid, 80, 124, 187
  anhydrous, 87
Citric acid cycle. *See* Krebs cycle
CL. *See* Chemically lean (CL)
"Classic" salami, 157
CO-containing tasteless smoke, 92
Coarse minced-mixed cooked salami, 194–195
Cochineal, 85
Coiled coil, 9
Cold
  shortening, 44–46
  smoke, 150, 184
Collagen, 9
  building blocks, 9
  casings, 106–107
  cross-links, 9
  elastin, 10
  solubility, 9–10
  triple helix of tropocollagen, 10f
Collagenases, 20
Color units (CU), 86
Color(s), 84–85
  betanin, 85
  Carmine, 85
  in cured meat products and fresh meat, 89
    color development mechanism, 96–99
    nitrate, 93–96
    nitrite, 93–96
    retention, 89–93
  enhancers, 99, 190, 200
    ascorbate, 100–101
    ascorbic acid, 100
    GDL, 101
    greening in cooked salami, 100
    nitrose gas, 100
    red curing color, 101
    in salami, 100
    sodium ascorbate, 100
  fermentation impact, 147
  fermented rice, 85–86
  groups, 85
  measurement, 101
  Paprika oleoresin, 86

Complex lipids, 21
Condensation water, 53–54
  formation, 53f
  temperatures of meat product and room temperatures, 54t
Condensation water, 197–198
Conditioning time, 144
Controlled spoilage, 111
Cooked salami, 61
Cooking, 196–197
Corium layer, 106–107
Cow meat, 114
CP. *See* Creatine phosphate (CP)
Creatine, 173–174
Creatine phosphate (CP), 17, 31
Crystal formation, 173–174
CU. *See* Color units (CU)
Cultures, 187–188
Cupping, 140–141
Cured meat products, 89, 93
  color development mechanism in, 96
    ascorbic acid, 98
    curing mechanism, 97–98, 98f
    fading of curing color, 99
    level of undissociated HNO_2, 97f
    nitrosometmyoglobin, 98
    NO, 96–97
    pH value, 97
    reasons, 98
    verdoheme complex, 99
  retention of color in, 89–93
Curing mechanism, 97–98, 98f
Cα. *See* Alpha-carbon atom (Cα)

D

"Dark" meat, 91
Deacarboxylation, 32–33
Deactivating peroxides, 78–79
Deamination, 33
Debaryomyces, 130
Debaryomyces hansenii (*D. hansenii*), 130
Defatted soy flour, 66
Degree of hardness of water, 73
Denaturation, 8
  of proteins, 49
Depot fat. *See* Subcutaneous fat
Dextrose, 178, 187–188
DFD meat. *See* Dry, firm, and dark meat (DFD meat)
DHA. *See* Docosahexaenoic acid (DHA)
Disaccharides, 68
Docosahexaenoic acid (DHA), 24

Dry, firm, and dark meat (DFD meat), 41–42, 91
  in comparison to normal meat postslaughter, 43f
  differences in certain characteristics, 43t
Dry ice, 135–136
Drying, 87, 117, 196–197
  materials, 103
  of salami, 162
    case hardening, 166–167
    impact of casing, 168f
    climatic parameters, 168–170
    distance and redirection water faces, 163f
    fermentation and, 164
    fresh products, 165
    fresh salami, 165f
    loss in weight during drying, 169f
    meat and fat particles, 163
    medium-fast–and slow-fermented salami, 167
    moisture, 166
    particle size of meat and fat, 170f
    QDS process, 170–171

E

Eh value, 52
  oxidation–reduction, 52f
Eicosapentaenoic acid (EPA), 24
Elastin, 10
Electromotive force. *See* Redox potential
Electron transfer chain (ETC), 35
Electroneutrality law, 52
Electrons, 52
Embden–Meyerhof pathway, 32, 127
Emulsifiers, 59
Encapsulated citric acid, 87
Endomysium, 10
Enterobacteriacea, 148
Enterobacteriaceae, 114, 146, 158
Enzymatic alkaline metabolic byproducts, 159
Enzyme-based reactions, 151
Enzymes in meat, 19–20
EPA. *See* Eicosapentaenoic acid (EPA)
Epimysium, 10
Erythorbate. *See* Ascorbate
Essential oils, 77–78
Ester, 21
ETC. *See* Electron transfer chain (ETC)
Extrusion techniques, 106–107

F

F-actin, 36–37
FAD. *See* Flavin adenine dinucleotide (FAD)

Fading of curing color, 99
Fast-acting starter cultures, 149
Fast-fermented salami, 154. *See also* Slow-fermented salami
  fermentation program for, 156t
  microflora in, 155f
Fast-fermented salamis, 154
Fats, 20–22, 117
  alpha-linolenic acid, 25f
  beef fat, 117–118
  building blocks, 21
  consumption, 3
  fatty acids, 21–23
  glycerol, 21
  groups, 21
  linoleic acid, 25f
  materials, 200–201
  and meat replacers in salami, 118–119
  melting points, 25–26
  molecule of triglyceride, 22f
  monounsaturated, 24
  oleic acid, 24f
  polyunsaturated fatty acids, 24–25
  rancidity, 27–30
  saturation, 23–24
  smearing, 160
  stearic acid, 22–23
  types, 23
  unsaturated fatty acids, 24, 26–27
Fatty acids, 21–23
  cis- or *trans*-configuration, 23f
  polyunsaturated, 24–25, 27
  unsaturated, 24, 26–27
Fermentation, 103, 141–142, 177
  air velocity, 146
  beef, 142–143
  and drying, 143
  drying room, 143–144
  fast-fermented salami, 154
  flavor in salami, 150–152
  German system, 142
  impact
    on color, 147
    on microbiological stability, 148
    on sliceability, 148–149
    on taste and aroma, 147–148
  medium-fast–fermented salami, 154–156
  Pseudomonas spp., 146
  salami manufacturing without fermentation room, 152
    alternating program, 152
    brine method, 152

salt method, 152
vacuum drying, 153–154
salami with noble mold, 160–162
slow-fermented salami, 157–160
small-diameter products, 144–145
smoking during fermentation, 150
Fermented rice, 85–86
Fermented salami. *See also* Non–heat treated fermented salami; Nonfermented salami (NFS)
pizza salami, 187–188
poultry salami, 188
semicooked and fully cooked fermented salami manufacturing technology, 185–186
summer sausage, 187
Fermented sausages, 111
FFAs. *See* Free fatty acids (FFAs)
Fibers, 87–88
Fibrous casings, 107–108, 139, 195
Filled salami, 152–153
Filled sausage, 178
Filling, 138–139, 195
casing material, 141
cupping, 140–141
fibrous casings, 139
filling speed, 139–140
modern filling machines, 140
process, 106
using wide filling horn, 141f
Finely cut cooked salami, 192–194
Flavin adenine dinucleotide (FAD), 32, 34–35
Flavor
enhancers, 71–72
of meat, 15–16
aldehydes, 16–17
boneless cuts of meat, 17
IMP, 16
PSE and DFD meat, 17
in salami, 150–152
Food
additives, 59
food-grade acid, 60
nutritional value of, 5–6
Free fatty acids (FFAs), 19–20, 28, 30
Free water, 12
Freeze-dried
meat, 116
starter cultures, 127
Freezer burning, 49
Freezing, 46–49
of ice crystals formation, 48f
speeds, 47
temperature curve by freezing meat, 47f

Fresh meat, 89
retention of color in, 89–93
Friction, 80–81
Frozen starter cultures. *See* Freeze-dried starter cultures
L-Fructose, 69f
Fuet, 183
Fully cooked fermented salami, 185
manufacturing technology, 185–186
Furanose, 69

G

Garlic (*Allium sativum*), 75–76
bacteriostatic properties of, 74
GDL. *See* Glucono-δ-lactone (GDL)
Gel made out of soy isolates, 67
Gelation of soy, 67–68
Genetically modified organism (GMO), 67
German salami, 179
German system, 141–142
Globulins, 11, 67
Gluconic acid, 86f, 185–186
Glucono-δ-lactone (GDL), 86, 86f, 101, 123–124
Glucose, 70, 124–125
α-Glucose, 70f
β-Glucose, 70f
D-Glucose, 69f
Glutaminic acid, 71
D-Glycerinaldehyde, 69f
Glycerol, 21
Glycine. *See* Alpha-carbon atom (Cα)
Glycogen, 32
Glycolysis, 31–33
Glycosidase, 20
GMO. *See* Genetically modified organism (GMO)
GMP. *See* 5′-Guanylate monophosphate (GMP)
Green pepper, 76
Greening in cooked salami, 100
GTP. *See* Guanosine triphosphate (GTP)
Guaiacol, 81–82
Guanosine triphosphate (GTP), 32, 34–35
5′-Guanylate monophosphate (GMP), 72

H

H_2O_2. *See* Hydrogen peroxide (H_2O_2)
Hardness, 73, 73t
Heat stabile spices, 77
Heat treatment, 200–202
Heavy metal ions, 80
Heme group, 89

Hemicellulose, 81
Hemoglobin, 11, 89
Hexanal, 30
Hexoses, 68
High protein solubility, 42–43
Higher-polymer phosphates, 60
Hog-casings, 105
Hot flavors, 75
Human eye, 101
Hungarian salami, 177
HVPs. *See* Hydrolyzed vegetable proteins (HVPs)
Hydrochloric acid (HCl), 78
Hydrogen atoms, 72–73
Hydrogen peroxide (H_2O_2), 20, 92–93, 99
Hydrogen sulfite (H_2S), 92–93
Hydrolytic rancidity, 28, 30
Hydrolyzed vegetable proteins (HVPs), 78
Hydroperoxides, 28
Hydroxyl groups, 4–5, 70
Hydroxyperoxides, 100

I

Ice, 72–73, 192
　crystals formation, 47, 48f
Identity-preserved program (IP program), 67
IEP. *See* Isoelectric point (IEP)
Immobilized water, 12
IMP. *See* 5′-Inosinate monophosphate (IMP); Inosine monophosphate (IMP)
5′-Inosinate monophosphate (IMP), 72
Inosine monophosphate (IMP), 16
Intermuscular fat, 21
Intramuscular fat, 21
IP program. *See* Identity-preserved program (IP program)
Isoelectric point (IEP), 12–14, 124
Italian system, 141–142

K

Kantwurst, 178–179
Ketone, 68
Ketose, 68, 69f
Krebs cycle, 33–35, 34f
　NADH, 33–34
　steps, 34

L

L*-A*-B* system, 101, 101f
LAB. *See* Lactic acid bacteria (LAB)

Lactic acid, 124, 148
　anaerobic glycolysis and formation of, 36–38
Lactic acid bacteria (LAB), 124–125, 148
Lactobacillus (*Lb.*), 99, 116, 123, 127
Lactose, 70, 71f, 124–125
Lb.. *See Lactobacillus* (*Lb.*)
Lean beef, 48
Lean meat, 113, 118
Lean muscle tissue, 3, 10–11
Ligands, 89
"Light" meat, 64, 91
Linear phosphates. *See* Chain–phosphates
Linoleic acid, 25, 25f
Lipases, 19–20
Lipids. *See* Fats
Lipolysis, 129
Liquid nitrogen, 126
Liquid smoke, 84. *See also* Natural smoke
Liquid-soluble spice flavorings, 77–78
Longer-chain phosphates in sausages, 62
Low-cost salami-type products, 103
Lup Cheong, 179–180

M

Maillard reaction, 54–55
Maltose, 124–125
D-Mannose, 69f
MAP packed. *See* Modified atmosphere packed (MAP packed)
Marbling fat. *See* Intramuscular fat
3-MCPD. *See* 3-Monochloropropane-1,2-diol (3-MCPD)
MDM. *See* Mechanically deboned meat (MDM)
Meat, 3, 37–38, 89, 112–113
　amino acids, 4–6
　Aw value, 51
　chicken MDM, 117
　collagen, 9–10
　Eh value, 52
　　oxidation–reduction, 52f
　enzymes in meat, 19–20
　flavor, 15–17
　freeze-dried meat, 116
　freezer burning, 49
　freezing, 46–49
　　of ice crystals formation, 48f
　　temperature curve by freezing meat, 47f
　lean meat, 113
　materials, 115, 190, 200–201
　muscle physiology, 10–15
　pH value, 49–51, 50f, 114

Index **209**

principles of muscle contraction and relaxation, 17–18
products, 89
proteins, 6–8
quality, 3
replacers in salami, 118–119
salami, 113
technology, 3, 4f
tempering, 48–49
thermal treatment, 116
trimmings, 43
Meat postslaughter, biochemical processes in, 36–38
anaerobic glycolysis and formation of lactic acid, 36–38
ATP levels postslaughter and rigor mortis, 38
Meat preslaughter, biochemical processes in, 31–36
ATP, 32f
formation, 31
rebuilding, 31
glycolysis, 32–33
krebs cycle, 33–35
oxidative phosphorylation, 35–36
source of energy, 31
Mechanically deboned meat (MDM), 201
Mechanically deboned meat from chicken (Chicken MDM), 117
Mechanically separated meat (MSM), 43–44
Medium-fast–fermented salami, 154–156, 167
decline in pH value in, 157f
2-Methoxy phenol. *See* Guaiacol
Metmyoglobin, 89–90
Microbiological stability, impact on, 148
Micrococcaceae, 128
Micrococcus spp., 125, 128–129
Middles, 105
Milano salami, 181–182
Minced materials, 137
Mincer-filling head
salami with, 137–138
system, 188
Mincer-mixing system, salami with, 136–137
Modified atmosphere packed (MAP packed), 171–172
fresh meat, 91
Moisture, 145–146, 166, 196
buffer, 106
Mold, 130, 160–161
prevention of unwanted, 131–132
Monascus angkak (*M. angkak*), 85–86
Monascus purpureus (*M. purpureus*), 85–86
3-Monochloropropane-1,2-diol (3-MCPD), 78

Monophosphates, 60
Monosaccharides, 68–69
Monosodium glutamate (MSG), 71–72
Monounsaturated fats, 24
MSG. *See* Monosodium glutamate (MSG)
MSM. *See* Mechanically separated meat (MSM)
Mucosa, 103, 105
Muscle(s), 10, 91
actin, 13, 14f
contractile unit of muscle fiber, 12
contraction and relaxation principles, 17
ATP, 17–18
binding of myosin to actin, 18f
relaxation of muscle after contraction, 18
relaxed and contracted sarcomere, 19f
sliding theory, 18
fiber contractile unit, 12
lean muscle tissue, 10–11
myosin, 12–13, 13f
pH value impact, 15f
protein-bound water, 12
sarcomere, 12f
structure, 11f
tissue, 31, 39
WHC, 13–14, 14f, 16f
Mustard-powder seed, 122
Mycotoxin, 131
Myofibrillar proteins, 11
Myoglobin, 11, 41, 89
different states, 90t
temporary acid denatures, 95
Myosin, 10–13, 13f, 31, 36, 38

N

NAD^+. *See* Nicotinamide adenine dinucleotide (NAD^+)
NADH dehydrogenase, 31
Natamycin, 70–71, 131–132
Natural casings, 103. *See also* Artificial casings
beef casings, 103–104
cap, 105
collagen casings, 106–107
filling process, 106
intestine, 104f
moisture buffer, 106
pig casings, 105
from sheep, 105–106
small intestines, 103
straight casings, 105
submucosa, 105

Natural nitrite, 96
Natural smoke, 80–81
   3,4-benzopyrene, 81
   functions, 82
   mistakes during smoking of salami, 83
   phenols, 81–82
   pryolysis, 81
   three methods of smoking, 82t
   wood, 81
   wood-chips or sawdust, 82
NFS. *See* Nonfermented salami (NFS)
NH_4^+. *See* Ammonium ion (NH_4^+)
Nicotinamide adenine dinucleotide (NAD^+), 31
Nitrate (NO_3), 93–96, 121
Nitric acid (HNO_3), 93
Nitric oxide (NO), 89, 96–97
Nitrite (NO_2), 93–96, 120–122, 146, 185, 187, 190, 199
   burn, 95
Nitrogen (N_2), 135–136
Nitrogen dioxide (NO_2), 95
Nitrosamines, 95
Nitrose gas (NO_2), 100, 190
Nitroso groups, 95
Nitrosometmyoglobin, 97–98
Nitrosomyoglobin, 185–186
Nitrous acid (HNO_2), 93, 158
Nitrous oxide (N_2O_3), 101
Noble mold, salami with, 160–162
Non-GMO products, 67–68
Nonenzymatic browning, 54–55
Nonfermented salami (NFS), 189. *See also* Fermented salami
   additives selection, 190–191
   Cabanossi, 200–201
   cheese salami, 199–200
   chicken salami, 201–202
   manufacturing technology, 191
      coarse minced-mixed cooked salami, 194–195
      cooking, 196–197
      drying, 196–197
      filling, 195
      finely cut cooked salami, 192–194
      manufacture of BE, 192
      packaging, 197–198
      precuring of meat and fat materials, 191–192
      smoking, 196–197
      storage, 197–198
   Polish salami, 199
   salami Florentine, 200
   selection and preparation of raw materials, 189–190
   Strasburg salami, 201
   Vienna salami, 200
Non–heat treated fermented salami. *See also* Nonfermented salami (NFS)
   fermentation of foods, 111
   manufacturing technology, 133
      drying of salami, 162–171
      fermentation, 141–162
      filling, 138–141
      packaging/storage, 171–174
      salami in bowl-cutter, 133–136
      slicing, 171
   selection of additives, 119–120
      acidification additives selection, 123–125
      case hardening, 122–123
      selected wanted mold and yeast, 130–132
      spices, 122
      starter cultures, 125–130
   selection of raw materials, 112
      fat, 117–119
      meat, 112–117
Nonwaterproof materials, 131
Nutritional value of food, 5–6

O

Oleic acid, 24, 24f
Oleoresins, 76–77
Oligosaccharides, 68
Omega-3 fatty acids, 24–25
Omega-6 fatty acids, 25
Onion (*Allium cepa*), 75–76
Oregano, 80
Overbreeding in pigs, 39
Oxidation, 27
Oxidative deterioration, 27
Oxidative phosphorylation, 35–36
Oxidative rancidity, 29
Oxymyoglobin, 89–91

P

P_2O_5 content of phosphates, 62
Packaging, 171–174, 197–198
Packed nonoxygenated fresh meat, 91
Paddle-mixer, 199, 201–202
PAH. *See* Polycyclic aromatic hydrocarbons (PAH)
Pale, soft and exudative meat (PSE meat), 40
   decline in pH value, 42, 42f
   differences in certain characteristics, 43t

Index   **211**

Paprika oleoresin, 76, 86
PCR test. *See* Polymerase chain reaction test (PCR test)
Pediococcus, 128
Penicilium, 130
  P. nalgiovense, 131
Pentoses, 68
Pepper, 76
Pepperoni, 183
Peptides, 7
Perimysium, 10
Peroxidation, 27
Peroxide value number (PV number), 29
pH. *See* Potential of hydrogen (pH)
Phenol radical, 79
Phenolic substances, 79
Phenols, 81–83
Phosphate-blend, 61–62
Phosphates, 59, 191
  P_2O_5 content, 62
  production and properties, 60
  properties, 60–62, 61f
Phosphoric acid, 60
Pig casings, 105
Pigs, 39
Pimaricin. *See* Natamycin
Pink pepper, 76
Pinkish touch, 94
Pizza salami, 183–184, 187–188
Polish salami, 199
Polyamides, 6–7
Polycyclic aromatic hydrocarbons (PAH), 80–81
Polyhydroxyketone. *See* Ketose
Polyhyrdoxyaldehyde. *See* Aldose
Polymerase chain reaction test (PCR test), 67
Polypeptides, 7
Polyphenols in green tea oil, 80
Polysaccharides, 81
Polyunsaturated fatty acids, 24–25
Pork, 178
  fat, 26–27, 117
  meat, 3, 40–41
  rind powder, 68
  runners, 103–104
  trimmings, 189
Post mortem glycolysis, 37–38
  formation of lactic acid, 37–38
Potassium chloride, 64–65
Potassium nitrate (KNO_3), 93
potentia hydrogeni. *See* Potential of hydrogen (pH)
Potential of hydrogen (pH), 49–51, 50f, 149

Poultry, 43
Poultry salami, 188
Precuring of meat and fat materials, 191–192
Predried meat, 115
Preslaughter muscle tissue, 31, 35–36
Primary structure of protein, 7
Procollagen, 9
Products in world
  beef salami, 184
  Cacciatore, 180
  Chorizo, 182
  Fuet, 183
  German salami, 179
  Hungarian salami, 177
  Kantwurst, 178–179
  Lup Cheong, 179–180
  Milano salami, 181–182
  Pepperoni, 183
  pizza salami, 183–184
  Sopressa, 181
  Sucuk, 182
Proteases, 20
Proteins, 6–7, 33
  denaturation, 8
  α-helix and β-pleated sheet secondary structure, 8f
  peptides, 7
  structures, 7–8
  total nitrogen, 8
Proteolysis, 20
Proteolytic mold, 131
PSE meat. see Pale. , soft and exudative meat (PSE meat)
Pseudomonas spp., 91–93, 146
  P. aeruginosa, 20
Pungent flavors, 75
PV number. *See* Peroxide value number (PV number)
Pyranose, 69
Pyrolysis, 80–81
Pyrophosphates, 60
Pyruvate, 32–33
Pyruvic acid, 75–76

Q

QDS process. *See* Quick-Dry-Slice process (QDS process)
Quality of meat and meat products, 3
Quaternary structure of protein, 8
Quick-Dry-Slice process (QDS process), 170–171

R

Radical, 27–28
Rancidity, 49, 151–152
  autoxidation, 27
  FFA content, 30
  hydrolytic, 30
  hydroperoxides, 28
  measurement ways, 29
  peroxidation, 27
  presence of oxygen, 27
  radical, 27–28
  TBA value, 29–30
  titration process, 29
  total oxidation volume, 30
Raw fermented salami, 111–113, 126, 142
Raw materials selection and preparation, 189–190
Red, soft, and exudative meat (RSE meat), 41
Red curing color, 101
Red muscles, 45
Redox potential, 52
Redox reaction couple, 52
Redox reactions. *See* Reduction–oxidation reactions (Redox reactions)
Reduced myoglobin, 89–90
Reduction–oxidation reactions (Redox reactions), 52
Refrigeration, 189, 197
Relative humidity (RH), 51, 53, 74, 112
Residual sugar, 149
RH. *See* Relative humidity (RH)
Ribonucleotide, 72
D-Ribose, 69f
Rigor mortis, 36, 41–42
  anaerobic glycolysis and formation of lactic acid, 36–38
  ATP levels postslaughter and, 38
  differences in meat preslaughter and postslaughter, 38t
Rigor shortening, 45
Ring and chain phosphates combinations, 60
Ring–phosphates, 60
Ripening, 38
Rosemary, 80
  extract, 121–122
Rosmanol, 80
Rounds, 103–105
Runners, 103–105

S

Sacaicin, 128
"Safety buffer", 149
Sage, 80
Salami, 103, 105–106, 119–120, 177, 189
  mass, 134
  raw fermented, 108
Salami Florentine, 200
Salami manufacturing without fermentation room, 152
  alternating program, 152
  brine method, 152
  salt method, 152
  vacuum drying, 153–154
Salmonella spp., 120
Salt, 119–122, 182, 185, 188, 190, 199
  method, 152
Salt–sodium chloride, 62–63
  functions in meat and meat products, 63
  increased levels of water, 64
  "light" meat products, 64
  pH values, 64
  impact of salt on protein, 63f
Sarcolemma, 10
Sarcomere, 12, 12f
Sarcoplasm (cytoplasm), 10
Sarcoplasmic proteins, 11
Sarcoplasmic reticulum (SR), 17, 44
Saturation of fat, 23–24
Sausage(s), 105–106
  mass, 181–182, 185
Sawdust, 82
Scoville heat units (SHUs), 75
Secondary amines, 95
Secondary structure of protein, 7
Semicooked fermented salami, 185
  manufacturing technology, 185–186
Semifrozen lean meat, 192–193
Serosa, 103, 105
Shelf-life, 76–78
Shirring, 106–107
Short-chain phosphates, 61
SHUs. *See* Scoville heat units (SHUs)
Six-carbon molecule and pyruvate, 32–33
Slaughtering process, 39
Sliceability, impact on, 148–149
Sliceable fermented salami, 111–112
Slicing, 171
Sliding theory, 18
Slow-fermented salami, 157. *See also* Fast-fermented salami
  caseinate, 159
  fat smearing, 160
  large-diameter products, 158–159
  natural acidification, 157–158

pH value, 158, 160f
uncontrolled fermentation, 159
Slow-fermented salami, 167
Small intestines, 103
Smoking, 80–81, 122, 196–197
    during fermentation, 150
Sodium ascorbate, 100
Sodium chloride (NaCl), 62–63
Sodium hydroxide (NaOH), 78
Sodium nitrite (NaNO$_2$), 93
Sodium tripolyphosphate (STPP), 59
Sodium-reduced sea salts, 65
Sol protein, 134
Sopressa, 181
Sorbate, 70–71
Sorbic acid, 70–71
Soy concentrate, 66
Soy protein, 66. *See also* Whey protein
    gelation of soy, 67–68
    production of soy products, 66
Spice blends, 74–75
Spices and extracts, 74
    bacteriostatic properties of garlic and thyme, 74
    carotenoids, 76
    essential oils, 77–78
    heat stabile spices, 77
    hot flavors, 75
    level of microbiological contamination in untreated spices, 76–77
    parts of plants, 75
    pepper, 76
    pyruvic acid, 75–76
    RH, 74
    spice blends, 74–75
    spice oleoresins or extracts, 77
    vanilla, 76
Spicy-hot materials, 122
Sprayed room, 132
Spraying process, 84
SR. *See* Sarcoplasmic reticulum (SR)
St. See Staphylococcus spp
Staphylococcus spp., 128–129
    S. aureus, 190, 197
Starter cultures, 125–126, 154
    fast starter cultures, 129–130
    freeze-dried, 127
    LAB, 128
    Lb. plantarum and *Pediococcus* spp., 127–128
    liquid nitrogen, 126
    Micrococcaceae, 128
    Micrococcus spp., 129

mixing process, 126–127
    Staphylococcus spp., 129
Steam condensation, 80–81
Steam-smoke, 80–81
Stearic acid, 22–23
L or D-Stereoisomers, 4
Steroids, 131–132
Storage, 171–174, 197–198
STPP. *See* Sodium tripolyphosphate (STPP)
Straight casings, 105
Strasburg salami, 201
Streaking, 82–83
Streptomyces natalensis (*S. natalensis*), 70–71
Subcutaneous fat, 21
Subcutaneous pork fat, 26
Sublimation, 49, 135–136
Submucosa, 103, 105
Sucking effect, 15, 87–88
Sucrose, 70, 71f, 124–125
Sucuk, 182
Sugars, 68
    for acidification and flavor, 124–125
    aldose, 68
    furanose, 69
    D-glucose, 69f
    D-Glycerinaldehyde, 69f
    hydroxyl groups, 70
    L-fructose, 69f
    pentoses, 68
    D-Ribose and D-mannose, 69f
    in salamis, 70
Summer sausage, 187
Surface configuration, 19
Sweet flavors, 75
Swiss Emmental, 200
Swollen material, 106–107
Syringol, 81–82

T

Tangy flavors, 75
Taste, fermentation impact on, 147–148
Tasteless smoke, 92
Taurin, 72
TBA value. *See* Thiobarbituric acid value (TBA value)
Temperature, 185–186
Tempering of meat, 48–49
Temporary hardness, 73
Tertiary structure of protein, 8
Thawing of meat, 48
Thermal treatment, 116, 190, 195, 201
Thiobarbituric acid value (TBA value), 29–30

Thyme, bacteriostatic properties of, 74
Tiger stripes, 82–83
Titration process, 29
α-Tocopherol, 79–80
β-Tocopherol, 79–80
γ-Tocopherol, 79–80
δ-Tocopherol, 79–80
Tocopherols, 79–80
Total nitrogen, 8
Total oxidation volume, 30
Total-plate-count (TPC), 74, 76–77
Tricarboxylic acid cycle (TCA). *See* Krebs cycle
Tropocollagen units, 9

U

"Umami" taste, 72
Uncured meat products. *See also* Cured meat products
  retention of color in, 89–93
Undissociated nitrous acid (HNO_2), 125, 147
Unsaturated fatty acids, 24, 26–27
Unwanted mold, prevention of, 131–132

V

Vacuum
  drying, 153–154
  packing products, 172
  vacuum-packed meat, 91
Vanilla, 76
Vanillin, 76
Vegetable extracts, 96

Verdoheme complex, 99
Vienna salami, 200
Visible particle fraction, 81–82

W

Water, 72–73
Water-holding capacity (WHC), 13–15, 40, 42–43
  due to addition of salt, 16f
  of DFD meat, 42–43
  of fibers, 87–88
  Maillard reaction, 54–55
  in relation to pH value, 14f
Waterproof casings, 103
Weasands, 103–104
WHC. *See* Water-holding capacity (WHC)
Whey protein, 65–66
Whilet potassium sorbate, 70–71
White pepper, 76
White-colored sausage, 119
Wood, 81
Wood-chips, 82
Woven textile casings, 108

Y

Yeast, 130, 160–161
  extracts, 78
  prevention of unwanted mold, 131–132

Z

Zwitterions, 6, 6f